KAM Stability and Celestial Mechanics

MEMOIRS
of the
American Mathematical Society

Number 878

KAM Stability and Celestial Mechanics

Alessandra Celletti
Luigi Chierchia

May 2007 • Volume 187 • Number 878 (third of four numbers) • ISSN 0065-9266

American Mathematical Society
Providence, Rhode Island

2000 *Mathematics Subject Classification.*
Primary 70F07, 70–04; Secondary 70H08, 37J40.

Library of Congress Cataloging-in-Publication Data

Celletti, A. (Alessandra)
KAM stability and celestial mechanics / Alessandra Celletti, Luigi Chierchia.
p. cm. — (Memoirs of the American Mathematical Society, ISSN 0065-9266 ; no. 878)
"May 2007, volume 187, number 878 (third of 4 numbers)."
Includes bibliographical references.
ISBN 978-0-8218-4169-3 (alk. paper)
1. Three-body problem. 2. Celestial mechanics. 3. Perturbation (Mathematics) I. Chierchia, Luigi, 1957– II. Title.

QB362.T5C45 2007
521—dc22 2007060667

Memoirs of the American Mathematical Society

This journal is devoted entirely to research in pure and applied mathematics.

Subscription information. The 2007 subscription begins with volume 185 and consists of six mailings, each containing one or more numbers. Subscription prices for 2007 are US$649 list, US$519 institutional member. A late charge of 10% of the subscription price will be imposed on orders received from nonmembers after January 1 of the subscription year. Subscribers outside the United States and India must pay a postage surcharge of US$38; subscribers in India must pay a postage surcharge of US$43. Expedited delivery to destinations in North America US$53; elsewhere US$130. Each number may be ordered separately; *please specify number* when ordering an individual number. For prices and titles of recently released numbers, see the New Publications sections of the *Notices of the American Mathematical Society.*

Back number information. For back issues see the *AMS Catalog of Publications.*

Subscriptions and orders should be addressed to the American Mathematical Society, P. O. Box 845904, Boston, MA 02284-5904, USA. *All orders must be accompanied by payment.* Other correspondence should be addressed to 201 Charles Street, Providence, RI 02904-2294, USA.

Memoirs of the American Mathematical Society is published bimonthly (each volume consisting usually of more than one number) by the American Mathematical Society at 201 Charles Street, Providence, RI 02904-2294, USA. Periodicals postage paid at Providence, RI. Postmaster: Send address changes to Memoirs, American Mathematical Society, 201 Charles Street, Providence, RI 02904-2294, USA.

This publication is indexed in *Science Citation Index*®, *SciSearch*®, *Research Alert*®, *CompuMath Citation Index*®, *Current Contents*®/*Physical, Chemical & Earth Sciences.*
Printed in the United States of America.

∞ The paper used in this book is acid-free and falls within the guidelines established to ensure permanence and durability.
Visit the AMS home page at http://www.ams.org/

10 9 8 7 6 5 4 3 2 1 12 11 10 09 08 07

Contents

Abstract

KAM theory is a powerful tool apt to prove perpetual stability in Hamiltonian systems, which are a perturbation of integrable ones. The smallness requirements for its applicability are well known to be extremely stringent. A long standing problem, in this context, is the application of KAM theory to "physical systems" for "observable" values of the perturbation parameters.

We consider the Restricted, Circular, Planar, Three-Body Problem (RCP3BP), i.e., the problem of studying the planar motions of a small body subject to the gravitational attraction of two primary bodies revolving on circular Keplerian orbits (which are assumed not to be influenced by the small body). When the mass ratio of the two primary bodies is small, the RCP3BP is described by a nearly-integrable Hamiltonian system with two degrees of freedom; in a region of phase space corresponding to nearly elliptical motions with non small eccentricities, the system is well described by Delaunay variables. The Sun-Jupiter observed motion is nearly circular and an asteroid of the Asteroidal belt may be assumed not to influence the Sun-Jupiter motion. The Jupiter-Sun mass ratio is slightly less than 1/1000.

We consider the motion of the asteroid 12 Victoria taking into account only the Sun-Jupiter gravitational attraction regarding such a system as a prototype of a RCP3BP. For values of mass ratios up to 1/1000, we prove the existence of two-dimensional KAM tori on a fixed three-dimensional energy level corresponding to the observed energy of the Sun-Jupiter-Victoria system. Such tori trap the evolution of phase points "close" to the observed physical data of the Sun-Jupiter-Victoria system. As a consequence, in the RCP3BP description, the motion of Victoria is proven to be forever close to an elliptical motion.

The proof is based on: 1) a new iso-energetic KAM theory; 2) an algorithm for computing iso-energetic, approximate Lindstedt series; 3) a computer-aided application of 1)+2) to the Sun-Jupiter-Victoria system.

The paper is self-contained but does not include the ($\sim$ 12000 lines) computer programs, which may be obtained by sending an e-mail to one of the authors.

Received by the editor December 2003 and in revised form May 2005.

2000 *Mathematics Subject Classification.* Primary 70F07, 70–04; Secondary 70H08, 37J40.

Key words and phrases. Three-body problems, Computer-aided proofs, KAM techniques.

CHAPTER 1

Introduction

1.1. Quasi-periodic solutions for the n-body problem

The n-body problem consists in studying the dynamics of n point masses in the three-dimensional space mutually attracted by Newton gravitational law. Such dynamics is governed by the following system of ordinary differential equations

$$\frac{d^2u^{(i)}}{dt^2} = - \sum_{\substack{0\leq j\leq n-1\\ j\neq i}} m_j \frac{u^{(i)} - u^{(j)}}{|u^{(i)} - u^{(j)}|^3} \ , \qquad i = 0, 1, ..., n-1 \ , \tag{1.1.1}$$

where m_j are the masses of the bodies, the n functions $t \in \mathbb{R} \to u^{(i)}(t) \in \mathbb{R}^3$ describe the position of the ith body in space at time t and $|u^{(i)} - u^{(j)}|$ is the Euclidean distance between $u^{(i)}$ and $u^{(j)}$; the gravitational constant has been set equal to one (which is always possible upon rescaling of time).

In this paper we will be mainly concerned with the construction of quasi-periodic solutions for (1.1.1) and with their dynamical relevance with particular emphasis on stability. Roughly speaking, a quasi-periodic solution of (1.1.1) is a solution, which, in suitable coordinates, may be expressed in terms of linear motions filling out a k-dimensional torus with $2 \leq k \leq d$, d being the number of degrees of freedom of the problem (we will come back later to a more precise definition of quasi-periodic solutions).

It is well known that for $n = 2$, equations (1.1.1) may be "explicitly" solved, while, for $n > 2$, despite the effort of mathematicians such as Kepler, Newton, Laplace, Lagrange, Weierstrass, Mittag-Leffler, Poincaré, Birkhoff, Siegel, Kolmogorov, Moser and Arnold, basic questions about the n-body problem are still unsolved. For example, there are no proofs of the existence of quasi-periodic solutions for the n body-problem with[1.1] $n \geq 4$.

Particularly relevant for Celestial Mechanics is the *planetary* case, i.e, the case when one of the bodies ("star") has mass significantly larger than the other bodies

[1.1] V.I. Arnold in 1963 ([**5**]) gave the first existence proof of quasi-periodic motions in the case of three co-planar bodies ($m_0 \gg m_1, m_2$) revolving on nearly-circular orbits; more than thirty years later, in 1995, Arnold's result was extended by Laskar and Robutel ([**91**], [**119**]) to the spatial three-body problem (small inclinations, small eccentricities). Other quasi-periodic solutions for the three-body problem (spanning "lower dimensional tori") where shown to exist by Jefferys and Moser [**76**] in 1966 (linearly unstable case) and recently by Féjoz [**54**] (linearly stable, planar) and by Biasco, Chierchia and Valdinoci [**15**] (linearly stable, spatial). A (long) proof of the existence of quasi-periodic solutions in the general n-body problem was announced in [**74**] by M. Herman in 1995, but his untimely death (2001) did not allow such - certainly beautiful - piece of work to be completed. **Added in proof:** after this paper was sumbitted, J. Féjoz did complete Herman's work on the existence of quasi-periodic motions for the general n-body problem in the (indeed beautiful) paper [**55**]. Another general existence result of quasi-periodic motions for the *planar* n-body problem has been established in [**16**].

("planets"):

$$m_0 \gg m_i \ , \qquad i = 1, ..., n-1 \ .$$

In such a case the n-body problem may be viewed as a *perturbation* of $(n-1)$ decoupled two-body problems (star-$i^{\rm th}$ planet, $i = 1, ..., n-1$). The perturbative approach was explored with great success, especially, by Poincaré, Birkhoff, Kolmogorov, Arnold and Moser. Poincaré, in particular, was awarded, by Oscar II, King of Sweden and Norway, a prize for a memoir on the 3-body problem[1.2]. Among many other things, in the fundamental work [**117**], Poincaré, following Newcombe and Lindstedt, tried to study the existence of quasi-periodic solutions for general Hamiltonian systems, by looking at power-series expansions in the perturbation parameter (*Lindstedt series*). Poincaré was not able to decide whether such series were convergent or not[1.3] and it was only nearly seventy years after that J. Moser ([**112**]) showed how KAM theory could be used in order to give an indirect proof of the convergence of Lindstedt series.

Kolmogorv-Arnold-Moser theory[1.4] represented a major breakthrough in the study of the behavior of general conservative dynamical systems, providing technical tools, based upon a quantitative "fast (Newton) iteration scheme" in suitable decreasing family of Banach function spaces, which allowed to overcome crucial difficulties related to resonances and small divisors. Moser, in [**112**], pointed out that the KAM construction of quasi-periodic solutions was uniform in complex parameters so that, by Weierstrass theorem on uniform limits of analytic functions, the parameter analyticity was inherited by the solutions. As a by-product the KAM quasi-periodic solutions are analytic in the parameter and hence (by the identity principle for holomorphic functions) Lindstedt series are convergent.

Direct proofs avoiding fast iteration methods are more than twenty years younger than KAM theory. The first (semi-)direct proof was given by H. Eliasson in a 1984 preprint (see, also, [**49**]) and it is based on an extension of the approach used by Siegel in 1942 ([**125**]) in order to prove the conjugacy of germs of analytic functions

---

[1.2] "Oscar II initiated a mathematical competition in 1887 to celebrate his sixtieth birthday in 1889. Poincaré was awarded the prize for a memoir he submitted on the 3-body problem in Celestial Mechanics. In this memoir Poincaré gave the first description of homoclinic points, gave the first mathematical description of chaotic motion, and was the first to make major use of the idea of invariant integrals. However, when the memoir was about to be published in Acta Mathematica, Phragmen, who was editing the memoir for publication, found an error. Poincaré realized that indeed he had made an error and Mittag-Leffler made strenuous efforts to prevent the publication of the incorrect version of the memoir. Between March 1887 and July 1890 Poincaré and Mittag-Leffler exchanged fifty letters mainly relating to the Birthday Competition, the first of these by Poincaré telling Mittag-Leffler that he intended to submit an entry, and of course the later of the 50 letters discusses the problem concerning the error. It is interesting that this error is now regarded as marking the birth of chaos theory. A revised version of Poincaré's memoir appeared in 1890." Extracted from an article by J.J. O'Connor and E.F. Robertson (http://www-gap.dcs.st-and.ac.uk/ history/Mathematicians/Poincare.html).

[1.3] "[· · ·] Il s'agit maintenant de reconnaître si ces séries sont convergentes. [· · ·] Il semble donc permis de conclure que le séries (2) ne convergent pas. Toutefois les raisonnement qui précède ne suffit pas pour établir ce point avec une rigueur complète. [· · ·] Ne peut-il pas arriver que le séries (2) convergent quand on donne aux x_i^o certaines valeurs convenablement choisies? [· · ·] Les raisonnements de ce Chapitre ne me permettent pas d'affirmer que ce fait ne se présentera pas. Tout ce qu'il m'est permis de dire, c'est qu'il est fort invraisemblable." [**117**], Tome II, Chapitre XIII, no. 146, 149.

[1.4] Fundamental references are: [**81**], [**4**], [**5**], [**110**], [**111**], [**112**], [**72**]. For texts discussing KAM theory and Celestial Mechanics, see, e.g., [**126**], [**113**] and [**6**].

to their linear part[1.5]. Explicit compensations for the coefficients of the Lindstedt series sufficient to yield convergence were proven in the 90's by Gallavotti, Gentile and Mastropietro ([**60**], [**61**], [**62**], [**63**], [**64**]) and by Chierchia and Falcolini ([**42**], [**43**]). Compensations in a different setting were also investigated in [**65**].

It has to be pointed out that the n-body problem does not fall in the category of "general Hamiltonian systems" to which KAM theory applies. In fact, typical n-body problems are strongly *degenerate* from the point of view of KAM theory[1.6] and this is the (technical) reason why - as mentioned above - basic results are still missing in the mathematical theory of the n-body problem.

However, in some special instances - such as the Restricted Circular Planar Three-Body Problem (RCP3BP, for short) - the non degeneracy assumptions required by KAM theory are indeed satisfied (the RCP3BP is iso-energetically non degenerate) and the existence of quasi-periodic solutions follows at once from standard KAM theory. A natural question is then:

can one establish the existence of quasi-periodic motions for the RCP3BP for observed values of the astronomical parameters entering into the problem?

Such question, to which in this paper will be given a positive answer, will be discussed in the coming sections.

1.2. A stability theorem for the Sun-Jupiter-Victoria system viewed as a restricted, circular, planar three-body problem

We now state our main result. By definition, the *restricted* three-body problem consists in studying the differential equation obtained from (1.1.1) by taking $n = 3$ and setting one of the masses, say m_2, equal to zero. In such a case, the equations for the "primary bodies" $(i = 0, 1)$ decouple from the equation for the position $u^{(2)}$ of the "minor" (or "small") body. In other words, the restricted problem consists in studying the motion of a minor body subject to the gravitational field generated by the motion of two primary bodies, which are assumed not to be influenced by the gravitational attraction of the minor body. The RCP3BP consists in assuming further that the relative motion of the primary bodies is *circular* and that the motion of the minor body takes place on the *plane* generated by the motion of the primaries.

[1.5] See also, [**73**], [**131**].

[1.6] A typical (classical) result in KAM theory may be formulated as follows. *Consider a nearly-integrable Hamiltonian system governed by a real-analytic Hamiltonian function* $H(x, y; \varepsilon) := H_0(y) + \varepsilon H_1(x, y)$ *defined on the phase space* $\mathcal{M} := \mathbb{T}^d \times B$ *(*B *being a ball in* $\mathbb{R}^d$*) endowed with the standard symplectic form* $\sum_{j=1}^d dx_j \wedge dy_j$. *Let* $y_0 \in B$ *be such that* $\omega := H_0'(y_0)$ *is a Diophantine vector (i.e., there* $\exists\ \gamma, \tau > 0$ *s.t.* $|\omega \cdot n| \geq \gamma |n|^{-\tau}$, $\forall\ n \in \mathbb{Z}^d \backslash \{0\}$*) and that* $\det H_0''(y_0) \neq 0$. *Then, there exists* $\varepsilon_0 > 0$ *such that the unperturbed torus* $\mathbb{T}^d \times \{y_0\}$ *can be analytically continued for* $|\varepsilon| < \varepsilon_0$ *into an invariant torus on which the* H*-flow is analytically conjugated to the linear flow* $\theta \in \mathbb{T}^d \to \theta + \omega t$. The condition on the Hessian matrix $H_0''(y_0)$ is sometimes referred to as non-degeneracy (or KAM non-degeneracy) condition. The non-degeneracy condition may be replaced by the so-called *iso-energetic non-degeneracy condition*, namely the requirement that $\det \begin{pmatrix} H_0''(y_0) & \omega \\ \omega & 0 \end{pmatrix} \neq 0$. In such a case, the unperturbed torus $\mathbb{T}^d \times \{y_0\}$ may be analytically continued within the *fixed energy level* $H^{-1}(H_0(y_0))$ in such a way that the H-flow is analytically conjugated to the linear flow $\theta \in \mathbb{T}^d \to \theta + (1 + a)\omega t$, a being a small real number depending (analytically) on ε. In the general planetary n-body problem, the Hamiltonian of the integrable limit, H_0, depends on y_j with $j < d$; thus the non-degeneracy conditions are violated in the whole phase space. See, e.g., [**6**]. A recent paper where the non-degeneracy conditions are weakened is [**122**].

As mentioned above, if m_1/m_0 is small, the problem becomes perturbative and can be described, in suitable physical units and using classical Delaunay action-angle variables, by a nearly-integrable Hamiltonian system with Hamiltonian

$$\begin{aligned} H(\ell,g,L,G;\varepsilon) &:= -\frac{1}{2L^2} - G + \varepsilon F(\ell,g,L,G;\varepsilon)\ , \\ (\ell,g,L,G)\in\mathcal{M} &:= \mathbb{T}^2\times\{0<G<L\}\ , \end{aligned}$$

where $\mathbb{T}^2 := \mathbb{R}^2/(2\pi\mathbb{Z}^2)$ is the standard 2-torus and the perturbation F is a real-analytic function on the phase space $\mathcal{M}$ endowed with the standard symplectic form $d\ell\wedge dL + dg\wedge dG$; here the constant distance between the primaries and their total mass (m_0+m_1) are normalized to one; ε is essentially m_1 (the mass of the smaller primary body) and the period of the circular motion of the primaries is normalized to 2π. The Delaunay variables are thoroughly discussed in a self-contained way in § 3.1÷3.3 below[1.7].

Next, we select a RCP3BP modeling the motion of an asteroid in our Solar System. As primary bodies, we consider the major bodies of our Solar System, namely, Sun and Jupiter. As small body we pick an asteroid from the Asteroidal Belt, namely, the asteroid 12 Victoria (the number refers to the standard classification of asteroidal objects; see, e.g., [**134**]). The observed astronomical data of Victoria are[1.8]

$$a_V \simeq 0.449\ , \qquad e_V =\simeq 0.220\ , \tag{1.2.1}$$

where a_V and e_V denote, respectively, the major semi-axis (divided by the major semi-axis of Jupiter) and the eccentricity of the osculating Keplerian ellipse on which Victoria is observed to move.

Considering the system Sun-Jupiter-Victoria as a RCP3BP involves, clearly, a lot of physical approximations: the Sun-Jupiter orbit is assumed to be circular (while the observed osculating orbit is a Keplerian ellipse of eccentricity $4.82\cdot10^{-2}$); the gravitational attraction of all other Solar System bodies are neglected (most notably, the attraction of Mars and Saturn); the orbit of Victoria is assumed to be co-planar with the Sun-Jupiter plane (while the observed relative inclination is about $1.961\cdot10^{-2}$); the shape and extension of the bodies are not taken into account (in particular, asteroids are typically far from being spherical and, therefore, far from

---

[1.7] A brief summary: denote by P_0 and P_1 the two primary bodies and by P_2 the minor body. Then:

(i) The angle ℓ is the mean-anomaly of the osculating ellipse associated to the two-body problem P_0-P_2 (the "major-minor body" or "Star-asteroid" system); $g=\gamma-\psi$, where γ is the argument of the perihelion of the osculating ellipse P_0-P_2 (with respect to an inertial, "heliocentric" frame with origin coinciding with the position of P_0), and ψ is the longitude of the "planet" P_1, that, having normalized the period of the motion of the primaries and their total masses to 1, coincides with the time t. The action variables are given by $L=\sqrt{a}/m_0^{1/6}$ and $G=L\sqrt{1-e^2}$, where a and e are, respectively, the semi-major axis and the eccentricity of the osculating P_0-P_2 ellipse.

(ii) The perturbative parameter ε is defined as $m_1/m_0^{2/3}$, which together with the normalization condition $m_0+m_1=1$ yields $m_1=\varepsilon-\frac{2}{3}\varepsilon^2+\frac{1}{3}\varepsilon^3+\cdots$, such series having radius of convergence greater than one.

(iii) Setting $x^{(i)}=u^{(i)}-u^{(0)}$, the perturbation F is the function $x^{(2)}\cdot x^{(1)}-|x^{(2)}-x^{(1)}|^{-1}$ expressed in terms of the Delaunay variables; $x^{(1)}$ being, simply, the relative circular motion of P_1: $x^{(1)}=\big(\cos(t_0+t),\sin(t_0+t)\big)$.

[1.8] The data are taken, also, from the NASA site [**134**]. The choice of Victoria is, obviously, rather arbitrary; we only remark that we did not want to consider the eccentricity as a further smallness parameter (comparable to the disregarded physical quantities; see below).

being well approximated by point masses); dissipative phenomena are neglected (tides, solar winds, Yarkovsky effect,...).

In considering Sun-Jupiter-Victoria as a three-body problem we make a further approximation, physically consistent with the already made approximations: *we replace the perturbation F with an ε-independent trigonometric polynomial H_1 of degree ten* obtained by expanding in power of

$$a := L^2 \qquad \text{and} \qquad e := \sqrt{1 - \frac{G^2}{L^2}} \tag{1.2.2}$$

the perturbation F and retaining those terms which are "quantitatively compatible" with the above approximations; the trigonometric polynomial thus obtained is:

$$\begin{aligned} H_1 \quad := \quad & -\Big(1 + \frac{a^2}{4} + \frac{9}{64}\,a^4 + \frac{3}{8}\,a^2 e^2\Big) + \Big(\frac{1}{2} + \frac{9}{16}a^2\Big)\,a^2 e\ \cos\ell \\ & -\Big(\frac{3}{8}\,a^3 + \frac{15}{64}a^5\Big)\ \cos(\ell + g) + \Big(\frac{9}{4} + \frac{5}{4}a^2\Big)\ a^2 e\ \cos(\ell + 2g) \\ & -\Big(\frac{3}{4}\,a^2 + \frac{5}{16}\,a^4\Big)\ \cos(2\,\ell + 2\,g) - \frac{3}{4}\,a^2 e\ \cos(3\,\ell + 2\,g) \\ & -\Big(\frac{5}{8}\,a^3 + \frac{35}{128}\,a^5\Big)\ \cos(3\,\ell + 3\,g) - \frac{35}{64}\,a^4\ \cos(4\,\ell + 4\,g) \\ & -\frac{63}{128}\,a^5\ \cos(5\,\ell + 5\,g)\ , \end{aligned} \tag{1.2.3}$$

where a and e are the action variable functions defined in (1.2.2).

Thus, the motion of the asteroid Victoria, in the framework of the RCP3BP, is governed by the Hamiltonian

$$\overline{H}_{\rm SJV}(\ell, g, L, G) := -\frac{1}{2L^2} - G + \varepsilon_{\rm SJ} H_1(\ell, g, L, G)\ , \qquad \varepsilon_{\rm SJ} := 0.954 \cdot 10^{-3}\ , \tag{1.2.4}$$

where the value $\varepsilon_{\rm SJ}$ is the observed value of the ratio of the masses of Jupiter and the Sun (in our system of units).

In view of (1.2.1) and (1.2.2) we *define* the osculating values of the actions L and G for the observed Victoria motion as

$$L_{\rm V} := 0.670 \simeq \sqrt{a_{\rm V}}\ , \qquad G_{\rm V} := 0.654 \simeq L_{\rm V}\sqrt{1 - e_{\rm V}^2}\ . \tag{1.2.5}$$

Finally, we also define the *energy value* associated to the observed Victoria motion as

$$\overline{E}_{\rm V} := -1.769\ ;$$

such value is obtained by adding to the Keplerian limiting energy $-\frac{1}{2L_{\rm V}^2} - G_{\rm V}$ the "secular" contribution given by $\varepsilon_{\rm SJ}$ times the average over the angles of H_1 evaluated on the osculating actions $L_{\rm V}$ and $G_{\rm V}$.

Denote by ϕ^t the flow governed by the Sun-Jupiter-Victoria Hamiltonian $\overline{H}_{\rm SJV}$ in (1.2.4)-(1.2.3), i.e., $(z_0, t) \in \mathcal{M} \times \mathbb{R} \to \phi^t(z_0) \in \mathcal{M}$ is the unique function satisfying the ODE

$$\frac{d}{dt}\phi^t(z_0) = J\,\nabla \overline{H}_{\rm SJV} \circ \phi^t(z_0)\ , \qquad \phi^0(z_0) = z_0\ ,$$

where J is the standard (4×4) symplectic matrix $\begin{pmatrix} 0 & I \\ -I & 0 \end{pmatrix}$ and where $z_0 := (\ell_0, g_0, L_0, G_0)$ is a point in the phase space $\mathcal{M}$. The main result of this paper may be shortly described as follows.

THEOREM 1.2.1. *One can construct, on the three-dimensional energy level*

$$\mathcal{S}_{\overline{E}_{\rm V}} = \left\{(\ell, g, L, G) \in \mathcal{M} : \ \overline{H}_{\rm SJV}(\ell, g, L, G) = \overline{E}_{\rm V}\right\} ,$$

a ϕ^t-invariant region $\mathcal{J}$, whose boundary is given by two KAM tori and whose interior contains the set

$$\left\{(\ell, g, L, G) \in \mathcal{S}_{\overline{E}_{\rm V}} : \ |L - L_{\rm V}| < 0.0015 \ , \qquad |G - G_{\rm V}| < 0.0059\right\} .$$

A similar statement holds if $\varepsilon_{\rm SJ}$ is replaced by any perturbative parameter ε with $|\varepsilon| \leq 10^{-3}$.

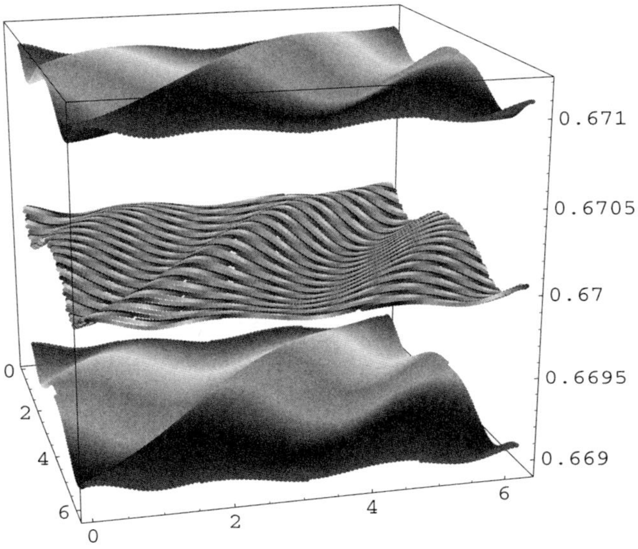

FIGURE 1. The energy level $\mathcal{S}_{\overline{E}_{\rm V}}$ in the chart $\{(\ell, g, L) \in \mathbb{T}^2 \times (0, \infty)\}$. The upper and lower surfaces are graphs within 10^{-9} from the invariant tori forming the boundary of $\mathcal{J}$; the intermediate surface is obtained integrating numerically a Sun-Jupiter-Victoria sample motion on the same energy level.

In other words, the evolution of astronomical data close to the astronomical observed data of Victoria, lie forever on nearly Keplerian orbits having major semi-axis and eccentricity close to the observed osculating data $a_{\rm V}$ and $e_{\rm V}$ of the asteroid Victoria.

The unabridged version of this statement is Theorem 4.4.1 at page 114.

1.3. About the proof of the Sun-Jupiter-Victoria stability theorem

The proof of Theorem 1.2.1 is based on a new computer implemented iso-energetic KAM theory, which allows to construct, on the fixed energy level $\mathcal{S}_{\overline{E}_{\rm V}}$, the two KAM tori forming the boundary of the invariant region $\mathcal{J}$.

The KAM method is designed so as to solve the parametric differential equations (quasi-linear PDE's on $\mathbb{T}^d$) in the symplectic phase space for the embedding functions of a KAM torus. No use of symplectic transformations is made, nor the system is required to be nearly-integrable: the method consists in providing a quantitative algorithm apt to detect solutions for the KAM tori equations in the vicinity of approximate solutions (on fixed energy levels). The method we present extends easily to general symplectic manifolds endowed with general symplectic forms. However, we refrain to present the theory in such generality having in mind the concrete application to the three body problem. The KAM method presented here is an evolution of the KAM theory worked out in § 4 of[1.9] **[25]**.

Roughly speaking, the basic idea is, essentially, the one beyond any constructive implicit function theorem: one starts with an approximate "non degenerate" solution and applies a theorem which guarantees that if the approximate solution is "good enough", then near-by there exists a (unique) true solution. In order to get effective quantitative estimates, it is important that the approximate solution is *not* the trivial one (while in standard KAM statements one constructs invariant tori as continuations of the unperturbed ones; see, for instance, the classical formulation in footnote 1.6).

Once the KAM machinery is worked out (this is done in § 2), the problem becomes to find "good" approximate solutions, to which the KAM theory of § 2 is applicable. In view of the appearance of resonances and small divisors, the problem of constructing approximate invariant tori, even from a numerical point of view, is well known to be a difficult one (see, e.g., **[68]**, **[67]**, **[18]**, **[41]**, **[75]**, **[127]**, **[130]**).

To describe the strategy followed in this paper, let us denote by $\theta \in \mathbb{T}^d \to Z^{\pm}(\theta) \in \mathcal{M}$ the embedding functions of the two KAM tori $\mathcal{T}^+$ and $\mathcal{T}^-$ forming the ("upper and lower") boundary of the invariant region $\mathcal{J}$. In order to find approximate solutions $Z^{\pm}_{\rm appr}$ to which apply the KAM Theorem 2.7.1 of § 2 so as to establish the existence of the KAM tori $Z^{\pm}$, we start by considering the iso-energetic Lindstedt series for two Diophantine tori close to the unperturbed tori corresponding to the Victoria osculating actions $L_{\rm V}$ and $G_{\rm V}$ in (1.2.5). We then construct explicitly the $12^{\rm th}$-order ε-truncation of such iso-energetic Lindstedt series and evaluate such truncations at the parameter value $\varepsilon = \varepsilon_{\rm SJ}$ corresponding to the Jupiter-Sun mass ratio. The thus obtained functions, which we shall denote here $Z^{\pm}_{\rm Lind}$, are trigonometric polynomials.

Even though in formulating the iso-energetic KAM Theorem 2.7.1 we take quite a lot of care in computing explicitly and carefully the various constants appearing in the statement[1.10], a direct application of the theorem to $Z^{\pm}_{\rm Lind}$ would not work: the KAM smallness requirements necessary in order to apply the KAM Theorem 2.7.1 are not met if one chooses as approximate solutions the functions $Z^{\pm}_{\rm Lind}$. To overcome this difficulty, using the Newton iteration scheme, which is at the basis of Theorem 2.7.1, we construct better approximate solutions, denoted here

[1.9] Analogous KAM techniques providing similar features in Lagrangian framework are older and go back to the "KAM theory in configuration space" of Salamon and Zehnder **[123]**, which, in turn, may be viewed as an evolution of the implicit-function method of E. Zehnder **[132]**, **[133]**. For an approach similar to that in **[25]** but more general and with a more geometrical insight see **[94]**.

[1.10] To get an idea of what we mean by that, take a look at Proposition 2.6.4, Lemma 2.7.2 and Lemma 2.7.4 below.

$Z^{\pm}_{\rm appr}$. The functions $Z^{\pm}_{\rm appr}$ are no more trigonometric polynomials (and, therefore, their construction is somewhat more implicit) but one has, nevertheless, an excellent control on their *norms*. The new approximations $Z^{\pm}_{\rm appr}$ do meet the smallness requirements of Theorem 2.7.1 and allow to construct the two invariant tori with embedding functions $Z^{\pm}$ forming the boundary of $\mathcal{J}$. The distance (in suitable analytic norms) of $Z^{\pm}_{\rm appr}$ from the final solutions $Z^{\pm}$ will be $\|Z^{\pm} - Z^{\pm}_{\rm appr}\| < 10^{-13}$ (compare with estimates (4.4.3), (4.4.4) below), while the distance of the solutions from the approximate Lindstedt polynomials $Z^{\pm}_{\rm Lind}$ satisfy $\|Z^{\pm} - Z^{\pm}_{\rm Lind}\| < 10^{-8}$.

The proof is *computer assisted*. As it is well known, it is possible to perform computations using a machine in such a way as to control the numerical errors introduced by the rounding-off performed by the machine; see, e.g., [**83**], [**48**], [**109**], [**84**], [**80**]. The idea is simple. Computers work with special classes of rational numbers, called "representable numbers". In general, an elementary operation (addition, subtraction, multiplication or division) between two representable numbers is no more a representable number. Therefore, computers perform a round-off of the result choosing "the closest" representable number. It is however, possible to give lower and upper bounds using representable numbers of the result of an elementary operation. In such a way the result of an elementary operation is no more a number but rather an *interval* (whose endpoints are representable numbers). One is then naturally led to substitute numbers with intervals and to perform elementary operations between such intervals of representable numbers. For example if a and b are real numbers, the operation $a + b$ will be replaced by the operation $[a_-, a_+] + [b_-, b_+] = [c_-, c_+]$ where $a_\pm$, $b_\pm$ and $c_\pm$ are representable numbers such that $a \in [a_-, a_+]$, $b \in [b_-, b_+]$ and c_- is a representable number smaller than $a_- + b_-$, while c_+ is a representable number greater than $a_+ + b_+$. For more information, see Appendix C or the above mentioned references.

The computer programs, which are used in the proofs are four: the first program deals with the choice of the "initial data" in phase space and their numerical properties; the second program (the main one) deals with the construction of the Lindstedt polynomials $Z^{\pm}_{\rm Lind}$; the third program evaluates the principal norms associated to $Z^{\pm}_{\rm Lind}$, while the fourth program estimates the norms of $Z^{\pm}_{\rm appr}$ and checks the applicability of the KAM Theorem 2.7.1 (for more details, see Appendix D).

The programs - available upon request to one of the authors - are not reproduced in the paper; however, all the algorithms and formulas used in the programs are thoroughly discussed in the text.

1.4. A short history of KAM stability estimates

As soon as KAM theory was formulated, astronomers tried to apply it to Celestial Mechanics. In 1966, the French astronomer M. Hénon [**70**] (see also [**71**]) pointed out that a *tout court* application of Arnold's theorem to the restricted three-body problem yields existence of invariant tori if the mass ratio of the primaries is less than[1.11] 10^{-333}; Hénon also noted that Moser's version gave better results since it applies for mass ratios up to 10^{-50}. These kind of results are so far from physical

[1.11] It appears as an accident that such number, 10^{-333}, is exactly the number of derivatives required by Moser in his first paper on KAM theory on the existence of invariant curves for smooth perturbations of integrable exact symplectic mappings of the annulus [**110**].

values[1.12] that astronomers and mathematicians begun to think that KAM was a beautiful "abstract" mathematical theory not really applicable to physics.

In the late 70's G. Gallavotti - also making KAM theory available at an undergraduate level (see, e.g., [**59**]) - gave a new impulse to this question and most improvements on KAM stability estimates came out under his direct or indirect influence.

The first improved KAM stability estimates concerned simple Hamiltonian models such as a forced pendulum or the so-called Chirikov-Greene ([**44**], [**66**]) standard map: see, in particular, [**29**], [**22**] (where computer-assisted techniques were introduced, for the first time, in the context of KAM estimates[1.13]), [**23**], [**118**], [**34**], [**23**], [**95**]. For more recent related results, see also [**35**].

In [**19**], [**20**] computer-assisted KAM estimates were established for the spin-orbit problem of Celestial Mechanics; see also [**27**] and [**21**] (where libration tori are constructed).

However, it is only in 1997 ([**25**]) that the problem raised by M. Hénon (i.e., to give KAM estimates for the restricted, three-body problem) was rigorously reconsidered. In [**25**] - with a strategy similar to the one used in this paper[1.14] - the existence of KAM tori for the RCP3BP for mass ratios of the primaries up to 10^{-6} was proved. KAM (computer-aided) estimates for the "secular part" of the Hamiltonian of the three-body problem are discussed in [**99**] and [**101**].

Much more abundant is the literature concerning non-rigorous methods for determining "optimal" existence estimates, i.e., parameter values above which a torus with a given frequency vector is expected to disappear[1.15] or to "break-down". Vaguely speaking, in Hamiltonian systems, the disappearance of KAM tori is related to the "onset" of a chaotic regime. A selection of articles - mostly related to the so-called "renormalization group" - on this subject is: [**44**], [**67**], [**51**], [**78**], [**124**], [**102**], [**9**], [**108**], [**50**], [**103**], [**115**], [**116**], [**7**], [**8**], [**53**], [**90**], [**106**], [**17**], [**87**], [**105**], [**30**], [**32**], [**57**], [**45**], [**82**], [**1**], [**31**], [**36**], [**58**], [**2**], [**38**], [**39**], [**79**], [**100**], [**40**], [**56**], [**26**], [**28**], [**37**].

While it goes beyond the scope of this introduction to review this literature, we want to make a couple of comments. One of the most reliable methods for determining the "break-down threshold" of KAM tori (at list in low dimensions) is the so-called Greene's *residue criterion* [**67**], which relates the break-down of tori with the change in stability of nearby periodic orbits. For partial rigorous justifications of Greene's criterion, see, e.g., [**52**], [**104**], [**47**]. Also quite effective seems to be

---

[1.12] The value 10^{-50} is about the proton-Sun mass ratio: the mass of the Sun is about $1.991 \cdot 10^{30}$ Kg, while the mass of a proton is about $1.672 \cdot 10^{-21}$ Kg, so that (mass of a proton)/(mass of the Sun) $\simeq 8.4 \cdot 10^{-52}$.

[1.13] Computers were also used in estimating the Siegel radius in [**92**] and [**93**].

[1.14] There are a lot of technical differences and a few conceptual ones between the approach used in [**25**] and the approach used here. As an example, we mention that in [**25**] we use standard non-degeneracy hypotheses (rather than *iso-energetical* non-degeneracy), since the point of that paper was just to show the possibility of constructing KAM tori in models not too far from "realistic ones". By the way, this point was overlooked in the first AMS Mathematical Review (MR1462771): the reviewer was, apparently, misled by the word "stability" used in the title (a new Review was posted in March 2001).

[1.15] For twist maps of the annulus there are theorems showing that above certain parameter values invariant curves do not exist any more; the first of this kind of theorems can be found in [**107**].

Laskar's *frequency analysis* method **[87]**, used in the numerical investigations of the behavior of the solar system (**[88]**, **[89]**). Another intriguing direction discovered numerically in **[10]** are the ε-complex properties of the conjugating function and the possible appearance of natural boundaries of singularities in ε-space; compare, e.g., with **[11]**, **[14]**, **[96]**, **[97]**, **[98]**, **[13]**, **[12]**.

A final side remark. The appearance of computers have changed quite radically science and life. In particular, astronomers that want to investigate the behavior of the Solar System use (more and more reliable) numerical simulations (see, e.g., **[3]**, **[85]**, **[114]**, **[128]**, **[86]**, **[88]**, **[89]**). We hope that this fact, based on the development of technology, does not subtract motivations for continuing the difficult mathematical investigations about the n-body problem, which still remains, in our opinion, one of the most rich and intriguing problems in mathematics.

1.5. A section-by-section summary

The paper is divided in four chapters, four appendices (and four above mentioned, not included computer programs). You are now reading the end of § 1 (the introduction). In § 2 the iso-energetic KAM theory is presented. In § 3, the classical Hamiltonian Delaunay theory for the restricted three-body problem is presented with full details. In § 4 the main stability theorem of the paper, concerning the perpetual stability of the asteroid Victoria in the framework of the RCP3BP Sun-Jupiter-Victoria, is stated (Theorem 4.4.1) and proved: the proof is divided into four main steps summarized at page 95.

More specifically:

- §2.1: Notations and a few conventions are introduced.
- §2.2: The definition of KAM torus is given and the partial differential equation satisfied by the embedding of a KAM torus on a fixed energy level is derived.
- §2.3: The notion of "approximate KAM torus" is introduced. The KAM algorithm is introduced and described at an "algebraic level". In particular the formulae which, given an approximate KAM torus, yield a new approximate torus, on the same energy level, are presented. The main result of this section is Proposition 2.3.5. The new approximate torus is constructed in such a way that the error term associated to it is (ignoring the effect of small divisors) quadratically smaller than the error associated to the initial approximation. The iso-energetical *non degeneracy* assumption is introduced; see Remark 2.3.10, where, in particular, the relations with the classical KAM non-degeneracy and iso-energetic non-degeneracy conditions are discussed.
- §2.4: The *KAM map* (i.e., the functional map which to an approximate torus associates the new approximate torus via the KAM algorithm described in § 2.3) is defined.
- §2.5: A few standard analytical technical tools, necessary to equip with estimates the KAM map, (including majorant theory and optimal small divisor estimates) are reviewed (and complete proofs are presented).
- §2.6: The KAM map is equipped with detailed and careful estimates. The results are summarized in Proposition 2.6.4. A suitable set of non negative numbers, say $\mathcal{N}$, controlling the relevant norms associated to an approximate torus is introduced; the algorithm, which to $\mathcal{N}$ associates $\mathcal{N}'$ (the set of non negative numbers controlling the corresponding norms associated to the new approximate torus); such algorithm is called the *KAM norm map*.
- §2.7: The main theorem in this KAM theory is presented: it is Theorem 2.7.1. Such theorem guarantees the existence, on a fixed energy level, of a KAM torus

nearby an approximate (iso-energetically non degenerate) KAM torus, provided the error term associated to the approximate torus is small enough, i.e., provided the "KAM smallness condition" (2.7.8) is satisfied. The numerical constants c_* and c_{**} appearing in the smallness condition (2.7.8) (together with a related constant $\hat{c}$) are explicitly evaluated in Lemma 2.7.4. The most technical part of this section is Lemma 2.7.2, on which the proof of Theorem 2.7.1 is based. The statement of Lemma 2.7.2 is a (lengthy!) list of simplified estimates, which allows to iterate infinitely many times the KAM map so as to construct the final KAM torus.

§2.8: The (analytic) dependence of KAM tori upon parameters is discussed. In particular the theory of Lindstedt series is briefly reviewed and the main recursive formulae, which allow to compute iso-energetic Lindstedt series are presented; see, in particular, Proposition 2.8.3.

§3.1: The Newton equations for the restricted three-body problem are introduced.

§3.2: The Delaunay theory for the Kepler two-body problem is reviewed. In particular, using Arnold-Liouville theorem, action-angle variables for the two-body problem are derived and physically interpreted.

§3.3: The full Hamiltonian for the restricted, circular, planar three-body problem is derived and expressed in terms of the Delaunay action-angle variables.

§3.4: The astronomical parameters associated to the Sun-Jupiter-Victoria system are introduced and a model, obtained truncating the perturbation function, is given. The physical criterion leading to this model is discussed in detail. The astronomically relevant phase space region associated to the model is also discussed.

§4.1: The iso-energetic Lindstedt series for the Sun-Jupiter-Asteroid problem are considered. The choice of the initial approximate tori, $Z^{\pm}_{\rm Lind}$, as the $12^{\rm th}$-order truncation of such series is discussed.

§4.2: The relevant norms associated to $Z^{\pm}_{\rm Lind}$ are computed.

§4.3: The KAM map is applied a few times (two times for $Z^{+}_{\rm Lind}$ and four times for $Z^{-}_{\rm Lind}$). The norms associated to the thus obtained new approximate tori $Z^{\pm}_{\rm appr}$ are evaluated.

§4.4: The KAM smallness condition of Theorem 2.7.1 is shown to be satisfied by $Z^{\pm}_{\rm appr}$. The main result on the perpetual stability of the motion of the asteroid Victoria in the framework of the Sun-Jupiter-Victoria problem, i.e., Theorem 4.4.1, is stated and proven.

A: The geometry of the ellipse is reviewed and a few classical formulae are derived.

B: Sharp small divisor estimates are explicitly performed. Diophantine constants for noble numbers are evaluated.

C: A short review of "interval arithmetic" and its use in computer-assisted proofs is reviewed.

D: A short guide to the computer programs used in the proof of Theorem 4.4.1 is presented.

Acknowledgments. We are indebted to P. Wittwer for many useful discussions and, especially, for very helpful suggestions about interval arithmetic. We thank E. Lega for interesting comments. We gratefully acknowledge the use of computers of the University "Roma Tre" and of the University of Roma "Tor Vergata".

This work has been supported by the MIUR (Ministero dell'Istruzione, dell'Università e della Ricerca) projects "*Dynamical Systems: Classical, Quantum, Stochastic*" and "*Variational Methods and Nonlinear Differential Equations*".

CHAPTER 2

Iso-energetic KAM Theory

2.1. Notations

Let $d \geq 2$ be an integer and consider a real-analytic *Hamiltonian function*

$$H = H(x, y) \ ,$$

where

$$x \in \mathbb{T}^d := \Big(\frac{\mathbb{R}}{2\pi\mathbb{Z}}\Big)^d$$

and $y \in \mathbb{R}^d$ are standard symplectic coordinates[2.1]. The Hamiltonian equations, with respect to the *standard symplectic form*

$$dx \wedge dy := \sum_{j=1}^{d} dx_j \wedge dy_j \ ,$$

read

$$\begin{aligned} \dot{x} &= H_y(x, y) \ , \\ \dot{y} &= -H_x(x, y) \ , \end{aligned} \tag{2.1.1}$$

where dot, as usual, denotes *time derivative*, and the subscripts denote, respectively, gradients with respect to the y or x variables. The flow generated by (2.1.1) will be denoted by

$$\phi_H^t(x, y) \ ,$$

which represents the solution $x(t), y(t)$ of (2.1.1) at time t with initial data at $t = 0$ given by $x(0) = x$ and $y(0) = y$.

The *inner product* between d-vectors a and b is denoted by

$$a \cdot b := \sum_{j=1}^{d} a_j b_j \ .$$

Given two vectors a and b, $a \otimes b$ denotes the matrix

$$(a \otimes b)_{ij} := a_i b_j \ .$$

If $\theta \in \mathbb{T}^d \to u(\theta) \in \mathbb{R}^d$ is a smooth function, u_θ denotes the Jacobian matrix with entries

$$\Big(u_\theta\Big)_{ij} := \frac{\partial u_i}{\partial \theta_j} \ .$$

[2.1] Even though the theory that we are going to discuss is valid for more general situations, for sake of concreteness and having in mind the application to the three-body problem, we shall confine ourselves to the case of $\mathbb{R}^d \times \mathbb{T}^d$ endowed with the standard simplectic form.

A superscript ^{T} denotes *matrix transposition*; A^{-T} denotes the transposed of the inverse of the square matrix A; the *unit* $(d \times d)$*-matrix* will be denoted by I or I_d if necessary.
In agreement with the above positions, H_{xx} or H_{xy} (and, analogously H_{yx} and H_{yy}) denote the $(d \times d)$-matrices

$$\Big(H_{xx}\Big)_{ij} := \frac{\partial^2 H}{\partial x_i \partial x_j} , \qquad \Big(H_{xy}\Big)_{ij} := \frac{\partial^2 H}{\partial x_i \partial y_j} ;$$

in particular

$$H_{yx} = H_{xy}^T .$$

Tensors of third derivatives are denoted as H_{xxx}, H_{yxx}, etc.; they may be seen as linear maps from d-vectors into $(d \times d)$-matrices: for example, for any $a \in \mathbb{C}^d$

$$\Big(H_{xyx} a\Big)_{ij} := \sum_{k=1}^{d} \frac{\partial^3 H}{\partial x_i \partial y_j \partial x_k} a_k ;$$

in particular, $H_{yxx} a = (H_{xyx} a)^T$.
If $\theta \in \mathbb{T}^d \to A(\theta) := (A_{ij})$ is a $(d \times d)$-matrix valued function, A_θ denotes the 3-tensor defined as follows: for $a \in \mathbb{C}^d$,

$$\Big(A_\theta a\Big)_{ij} := \sum_{k=1}^{d} \frac{\partial A_{ik}}{\partial \theta_j} a_k .$$

Average over $\mathbb{T}^d$ is denoted by $\langle \cdot \rangle$:

$$\langle u \rangle := \int_{\mathbb{T}^d} u(\theta) \frac{d\theta}{(2\pi)^d} .$$

Given a vector $\omega \in \mathbb{R}^d$, D_ω will denote the directional derivative on $\mathbb{T}^d$ given by

$$D_\omega := \omega \cdot \partial_\theta := \sum_{j=1}^{d} \omega_j \frac{\partial}{\partial \theta_j} . \tag{2.1.2}$$

A vector $\omega \in \mathbb{R}^d$ will be called *Diophantine* if there exist $\gamma > 0$, $\tau \geq d - 1$ such that

$$|\omega \cdot n| = \Big| \sum_{j=1}^{d} \omega_j n_j \Big| \geq \frac{\gamma}{|n|^\tau} , \qquad \forall\, n \in \mathbb{Z}^d \backslash \{0\} , \tag{2.1.3}$$

where $|\cdot|$ denotes the usual Euclidean 2-norm

$$|a| := \sum_{j=1}^{d} \sqrt{|a_j|^2} .$$

Given a Diophantine vector and a (vector or matrix valued) real-analytic function $\theta \in \mathbb{T}^d \to u(\theta)$ with vanishing average, $D_\omega^{-1} u$ denotes the real-analytic function[2.2]

$$D_\omega^{-1} u(\theta) := \sum_{\substack{n \in \mathbb{Z}^d \\ n \neq 0}} \frac{u_n}{i\omega \cdot n} \exp(in \cdot \theta) ,$$

[2.2] The analyticity of $D_\omega^{-1} u$ is consequence of the exponential decay of Fourier coefficients of analytic functions and of the Diophantine assumption on ω.

where u_n denotes Fourier coefficients,

$$u_n := \frac{1}{2\pi^d} \int_{\mathbb{T}^d} u(\theta)\ \exp(-in \cdot \theta) d\theta \quad ,$$

and i denotes $\sqrt{-1}$. In other words, $D_\omega^{-1} u$ is the unique analytic solution v of

$$D_\omega v = u \ , \qquad \langle v \rangle = 0 \ .$$

Finally, in order to simplify a little bit the notations, we shall not distinguish between "row-vectors" and "column-vectors", since the right interpretation will be clear from the context: for example, if A is a $(d \times d)$-matrix and $a \in \mathbb{C}^d$, then aA denotes the vector with k^{th} component $\sum_{j=1}^d a_j A_{jk}$, while Aa denotes the vector with k^{th} component $\sum_{j=1}^d A_{kj} a_j$.

2.2. KAM tori

Given a Diophantine vector $\omega \in \mathbb{R}^d$, a **KAM torus with frequency** ω is an H-invariant (i.e., invariant for the flow ϕ_H^t) surface embedded in the phase space $\mathbb{T}^d \times \mathbb{R}^d$, described parametrically, for $\theta \in \mathbb{T}^d$, by

$$\begin{aligned} x(\theta) &= \theta + \tilde{u}(\theta) \ , \\ y(\theta) &= \tilde{v}(\theta) \ , \end{aligned} \tag{2.2.1}$$

where $\tilde{u}$, $\tilde{v}$ are real-analytic functions defined on $\mathbb{T}^d$ and

$$\det\Big(I + \tilde{u}_\theta(\theta)\Big) \neq 0 \ , \qquad \forall\ \theta \in \mathbb{T}^d \ ; \tag{2.2.2}$$

furthermore the H-flow in the θ coordinate is required to be linear, i.e.:

$$\phi_H^t\Big(x(\theta), y(\theta)\Big) = \Big(x(\theta + \omega t), y(\theta + \omega t)\Big) \ . \tag{2.2.3}$$

Inserting $(x(\theta), y(\theta))$ into the Hamilton equations (2.1.1), in view of the rational independence of ω, one obtains that the functions $\tilde{u}$ and $\tilde{v}$ satisfy the following quasi-linear system of PDE's on $\mathbb{T}^d$:

$$\begin{aligned} \omega + D_\omega \tilde{u} - H_y(\theta + \tilde{u}, \tilde{v}) &= 0 \ , \\ D_\omega \tilde{v} + H_x(\theta + \tilde{u}, \tilde{v}) &= 0 \ , \end{aligned} \tag{2.2.4}$$

where D_ω is defined in (2.1.2). Without loss of generality, we may assume that

$$\tilde{u}(0) = 0 \ , \tag{2.2.5}$$

(which is a "normalization" condition that amounts to fix in $\theta = 0$ the "origin" of the invariant torus).

Vice-versa, given a triple $(\tilde{u}, \tilde{v}, \omega)$ satisfying (2.2.4) and (2.2.2), one obtains, via (2.2.1), a ϕ_H^t-invariant torus satisfying (2.2.3). By slight abuse of notation, *we shall also call the triple $(\tilde{u}, \tilde{v}, \omega)$ a KAM torus with frequency ω.*

We shall be particularly interested in *KAM tori lying over a given energy surface $H^{-1}(E)$*; in such a case (in view of (2.2.5)), one has

$$H(0, \tilde{v}(0)) = E \ .$$

Thus, the system of equations that a KAM torus with frequency ω lying on the energy level $H^{-1}(E)$ satisfies is[2.3]

$$\begin{aligned} \omega + D_\omega \tilde{u} - H_y(\theta + \tilde{u}, \tilde{v}) &= 0 , \\ D_\omega \tilde{v} + H_x(\theta + \tilde{u}, \tilde{v}) &= 0 , \\ \tilde{u}(0) &= 0 , \\ H(0, \tilde{v}(0)) - E &= 0 . \end{aligned} \tag{2.2.6}$$

Henceforth, we shall fix an energy level $E \in \mathbb{R}$, and discuss how to construct KAM tori on the fixed energy surface $H^{-1}(E)$.

2.3. Newton scheme for finding iso-energetic KAM tori

The starting point of an iterative method for finding solutions of (2.2.6) is the notion of *approximate KAM torus.* By definition, an **approximate KAM torus** *is a triple (u, v, ω) where u and v are real-analytic $\mathbb{R}^d$-valued functions on $\mathbb{T}^d$ with*

$$\det \mathcal{M}(\theta) := \det \big(I + u_\theta \big) \neq 0 , \quad \forall\, \theta \in \mathbb{T}^d , \qquad u(0) = 0 ,$$

and ω is a Diophantine vector (i.e., verifies (2.1.3)). To an approximate KAM torus (and the fixed energy level E) we associate the system

$$\begin{aligned} \omega + D_\omega u - H_y(\theta + u, v) &= f , \\ D_\omega v + H_x(\theta + u, v) &= g , \\ u(0) &= 0 , \\ H(0, v(0)) - E &= h , \end{aligned} \tag{2.3.1}$$

which *defines* the "error functions" f, g and the "error number" h. Obviously, if the error functions f and g and the number h vanish, the approximate KAM torus *defines* a KAM torus with frequency ω and energy E. The idea beyond the Newton iteration is that, if the error functions f and g and the number h are *small enough*, then one can start an iterative process leading to *better and better approximate KAM tori* (meaning that the new errors become nearly *quadratically* smaller).

In this paragraph we shall describe, *at an algebraic level*, how to construct a new approximate KAM torus given by

$$u' := u + z , \qquad v' := v + w , \qquad \omega' := (1 + a)\omega \;=: \omega_a ,$$

starting from an approximate KAM torus (u, v, ω). Notice, that, for $a \neq -1$, ω_a is automatically Diophantine with Diophantine constants $\gamma' := |1 + a|\ \gamma$ and τ. Varying ω by a (small) factor will be necessary in order to meet the energy constraint, as explained below.

NOTATIONAL REMARK 2.3.1. In what follows, given an approximate KAM torus (u, v, ω), H^0 (or H^0_x, etc.) will be short for $H(\theta + u(\theta), v(\theta))$ (or for $H_x(\theta + u(\theta), v(\theta))$, etc.).

We start with a preliminary result, which reflects the symplectic structure of the problem.

[2.3] Clearly, if ω is rationally independent and u and v satisfy the differential equations $\omega + D_\omega u - H_y(\theta + u, v) = 0$ and $D_\omega v + H_x(\theta + u, v) = 0$, then $H\big(\theta + u(\theta), v(\theta)\big)$ is constant in $\theta \in \mathbb{T}^d$. In fact, the differential equations imply that the function $t \to H\big(\omega t + u(\omega t), v(\omega t)\big)$ is constant in t (as one checks by differentiating with respect to t) and the density of the flow $t \to \omega t$ on $\mathbb{T}^d$ yields the claim.

LEMMA 2.3.2. *Let (u, v, ω) be an approximate KAM torus and let f and g be as in (2.3.1). Then $\mathcal{M} := I + u_\theta$ and v_θ verify*

$$\begin{aligned} D_\omega \mathcal{M} - H^0_{yx}\mathcal{M} - H^0_{yy} v_\theta &= f_\theta \ , \\ D_\omega v_\theta + H^0_{xx}\mathcal{M} + H^0_{xy} v_\theta &= g_\theta \ . \end{aligned} \tag{2.3.2}$$

If one defines

$$\begin{aligned} b &:= v_\theta^T f - \mathcal{M}^T g \\ \mathcal{B} &:= \mathcal{M}^T v_\theta - v_\theta^T \mathcal{M} \ , \\ \mathcal{G} &:= f_\theta^T v_\theta + \mathcal{M}^T g_\theta - v_\theta^T f_\theta - g_\theta^T \mathcal{M} \ , \end{aligned} \tag{2.3.3}$$

then,

$$\langle b \rangle = 0 \ , \qquad \langle \mathcal{B} \rangle = 0 \ , \tag{2.3.4}$$

and $\mathcal{B}$ satisfies the equation

$$D_\omega \mathcal{B} = \mathcal{G} \ , \tag{2.3.5}$$

so that, in particular,

$$\langle \mathcal{G} \rangle = 0 \ .$$

PROOF. The system of equations (2.3.2) is obtained immediately by taking the gradient with respect to θ of the first two equations in (2.3.1).

The following identities are immediate consequence of the given definitions:

$$v_\theta^T \omega = D_\omega v \ , \tag{2.3.6}$$

$$\partial_\theta H^0 = \mathcal{M}^T H^0_x + v_\theta^T H^0_y \ . \tag{2.3.7}$$

Thus,

$$\begin{aligned} \langle b \rangle &:= \langle v_\theta^T f - \mathcal{M}^T g \rangle \\ &\overset{(2.3.1)}{=} \langle v_\theta^T \omega + v_\theta^T D_\omega u - v_\theta^T H^0_y - \mathcal{M}^T D_\omega v - \mathcal{M}^T H^0_x \rangle \\ &\overset{(2.3.6)}{=} \langle v_\theta^T D_\omega u - u_\theta^T D_\omega v \rangle - \langle v_\theta^T H^0_y + \mathcal{M}^T H^0_x \rangle \\ &\overset{(2.3.7)}{=} \langle v_\theta^T D_\omega u - u_\theta^T D_\omega v \rangle \\ &= 0 \ , \end{aligned}$$

where the last identity follows by a double integration by parts[2.4]. This proves the first equation in (2.3.4).

The second equation in (2.3.4) also follows immediately by a double integration by parts[2.5]:

$$\langle \mathcal{B} \rangle := \langle \mathcal{M}^T v_\theta - v_\theta^T \mathcal{M} \rangle = \langle u_\theta^T v_\theta - v_\theta^T u_\theta \rangle = 0 \ .$$

From (2.3.2) and its transposed version, (2.3.5) follows at once. □

[2.4] In fact: $\displaystyle\int_{\mathbb{T}^d} \frac{\partial v_k}{\partial \theta_i} D_\omega u_k = \int_{\mathbb{T}^d} (D_\omega v_k) \frac{\partial u_k}{\partial \theta_i}$.

[2.5] In fact: $\displaystyle\int_{\mathbb{T}^d} \frac{\partial v_k}{\partial \theta_i} \frac{\partial u_k}{\partial \theta_j} = \int_{\mathbb{T}^d} \frac{\partial v_k}{\partial \theta_j} \frac{\partial u_k}{\partial \theta_i}$.

REMARK 2.3.3. If (u, v, ω) is an approximate KAM torus, then so is $(u, v, \omega_a) = (u, v, (1+a)\omega)$ with associated system given by

$$\begin{aligned} \omega_a + D_{\omega_a} u - H^0_y &= f_a \ , \\ D_{\omega_a} v + H^0_x &= g_a \ , \\ u(0) &= 0 \ , \\ H(0, v(0)) - E &= h \ ; \end{aligned} \tag{2.3.8}$$

the relation between the pair (f_a, g_a) defined in (2.3.8) and the pair (f, g) defined in (2.3.1) is given by

$$\begin{aligned} f_a &:= (1+a)f + a\ H^0_y \ , \\ g_a &:= (1+a)g - a\ H^0_x \ . \end{aligned} \tag{2.3.9}$$

Clearly Lemma 2.3.2 holds also for (u, v, ω_a) (for any a) with the obvious changes. For sake of clarity, we now translate Lemma 2.3.2 in the a-dependent case.

LEMMA 2.3.4. *Let (u, v, ω_a) be an approximate KAM torus. Then $\mathcal{M} := I + u_\theta$ and v_θ verify*

$$\begin{aligned} D_{\omega_a}\mathcal{M} - H^0_{yx}\mathcal{M} - H^0_{yy} v_\theta &= f_{a,\theta} \ , \\ D_{\omega_a} v_\theta + H^0_{xx}\mathcal{M} + H^0_{xy} v_\theta &= g_{a,\theta} \ . \end{aligned} \tag{2.3.10}$$

If one defines

$$\begin{aligned} & b_a := v_\theta^T f_a - \mathcal{M}^T g_a \\ & \mathcal{B} := \mathcal{M}^T v_\theta - v_\theta^T \mathcal{M} \ , \\ & \mathcal{G}_a := f^T_{a,\theta} v_\theta + \mathcal{M}^T g_{a,\theta} - v_\theta^T f_{a,\theta} - g^T_{a,\theta}\mathcal{M} \ , \end{aligned} \tag{2.3.11}$$

then,

$$\langle b_a \rangle = 0 \ , \qquad \langle \mathcal{B} \rangle = 0 \ ,$$

and $\mathcal{B}$ satisfies the equation

$$D_{\omega_a}\mathcal{B} = \mathcal{G}_a \ , \tag{2.3.12}$$

so that, in particular,

$$\langle \mathcal{G}_a \rangle = 0 \ .$$

The Newton-KAM algorithm is described in the following

PROPOSITION 2.3.5. *Fix $E \in \mathbb{R}$ and let (u, v, ω) be an approximate KAM torus. Define*[2.6]

$$\mathcal{T} := \mathcal{M}^{-1} H^0_{yy} \mathcal{M}^{-T} \ , \qquad \mathcal{M} := I + u_\theta \ , \tag{2.3.13}$$

and let, as above,

$$\begin{aligned} & f_a := (1+a)f + a\ H^0_y \ , \qquad g_a := (1+a)g - aH^0_x \ , \\ & b_a := v_\theta^T f_a - \mathcal{M}^T g_a \ , \qquad \mathcal{G}_a := f^T_{a,\theta} v_\theta + \mathcal{M}^T g_{a,\theta} - v_\theta^T f_{a,\theta} - g^T_{a,\theta}\mathcal{M} \ . \end{aligned}$$

[2.6]The suffix $H^0_{\cdot}$ means that the argument of $H_{\cdot}$ is $(\theta + u(\theta), v(\theta))$.

By Lemma 2.3.4, $\langle b_a \rangle = 0$ *and* $\langle \mathcal{G}_a \rangle = 0$. *Assume, now, that there exist* $c \in \mathbb{R}^d$ *and* $a \in \mathbb{R}\backslash\{-1\}$ *such that*

$$\begin{cases} \langle \mathcal{T} \rangle c + \langle \mathcal{T} D_{\omega_a}^{-1} b_a \rangle - \langle \mathcal{M}^{-1} f_a \rangle = 0 , \\ H\Big(0, v(0) + \mathcal{M}^{-T}(0)\left[c + D_{\omega_a}^{-1} b_a(0)\right]\Big) = E , \end{cases} \tag{2.3.14}$$

where $\omega_a := (1+a)\omega$, $D_{\omega_a} := \omega_a \cdot \partial_\theta$. *Define*

$$\begin{aligned} \hat{z} &:= \mathcal{T} c + \mathcal{T} D_{\omega_a}^{-1} b_a - \mathcal{M}^{-1} f_a , \\ \hat{z}_0 &:= -\big(D_{\omega_a}^{-1} \hat{z}\big)(0) , \\ z &:= \mathcal{M} D_{\omega_a}^{-1} \hat{z} + \mathcal{M} \hat{z}_0 , \\ w &:= \mathcal{M}^{-T}(v_\theta^T z + c + D_{\omega_a}^{-1} b_a) , \end{aligned} \tag{2.3.15}$$

and assume that

$$\det\big(\mathcal{M} + z_\theta\big) \neq 0 .$$

Then, the triple

$$(u', v', \omega') := (u + z, v + w, \omega_a) \tag{2.3.16}$$

is an approximate KAM torus satisfying

$$\begin{aligned} \omega_a + D_{\omega_a} u' - H_y(\theta + u', v') &= f' , \\ D_{\omega_a} v' + H_x(\theta + u', v') &= g' , \\ u'(0) &= 0 , \\ H(0, v'(0)) - E &= 0 , \end{aligned} \tag{2.3.17}$$

where f' *and* g' *verify the following relations. Let*

$$\begin{aligned} Q_1 &:= -[H_y(\theta + u + z, v + w) - H_y^0 - H_{yx}^0 z - H_{yy}^0 w] , \\ Q_2 &:= f_{a,\theta} \mathcal{M}^{-1} z , \\ Q_3 &:= H_{yy}^0 \mathcal{M}^{-T}(D_{\omega_a}^{-1} \mathcal{G}_a)\, \mathcal{M}^{-1} z , \\ Q_4 &:= H_x(\theta + u + z, v + w) - H_x^0 - H_{xx}^0 z - H_{xy}^0 w , \\ Q_5 &:= \mathcal{M}^{-T} g_{a,\theta}^T z , \\ Q_6 &:= -\mathcal{M}^{-T} f_{a,\theta}^T w , \\ Q_7 &:= \mathcal{M}^{-T}\, v_\theta^T (Q_2 + Q_3) , \end{aligned}$$

then

$$f' := Q_1 + Q_2 + Q_3 , \qquad g' := Q_4 + Q_5 + Q_6 + Q_7 .$$

REMARK 2.3.6. (i) The meaning of (2.3.14) is the following. The first relation in (2.3.14) is equivalent to require that $\langle \hat{z} \rangle = 0$, which is necessary in order to be able to define $D_{\omega_a}^{-1} \hat{z}$ and (hence) z.
As for the second relation in (2.3.14), observe that

$$v'(0) = v(0) + \mathcal{M}^{-T}(0)[c + D_{\omega_a}^{-1} b_a(0)] ,$$

which is exactly the second argument of H in the second equation of (2.3.14). Therefore, the second relation in (2.3.14) is equivalent to last equation in (2.3.17). In particular, this means that the new approximate torus describes a surface which sits *exactly* on the energy level $H^{-1}(E)$. The solvability of equations (2.3.14) will be discussed below (see, in particular, Lemma 2.3.9); we only observe, here, that, if the matrix $\langle \mathcal{T} \rangle$ is invertible (which, in general, is not true), one can express c in terms of a and (2.3.14) reduces to a single equation for the scalar unknown a.

(ii) The definition of $\hat{z}_0$ implies that $z(0) = 0$. This implies immediately the third equation in (2.3.17).

(iii) The quadratic character of the iteration scheme described in the above proposition may be easily explained as follows. Imagine to replace the error functions f and g, respectively, by ηf and ηg with some small parameter η. Now, assume that

$$|c| \ , \ |a| \sim \eta \ . \tag{2.3.18}$$

Then, by (2.3.9), it is $f_a = O(\eta) = g_a$. Then, also

$$b_a = O(\eta) \ , \qquad \mathcal{G}_a = O(\eta) \ . \tag{2.3.19}$$

The relations (2.3.18) and (2.3.19) imply, now, that

$$z = O(\eta) \ , \qquad w = O(\eta) \ ,$$

which, in turn, yields (as it is immediate to check)

$$Q_i = O(\eta^2) \ .$$

But this means that $f' = O(\eta^2) = g'$.

PROOF. (of Proposition 2.3.5) We start by noticing that (2.3.10) may be rewritten as

$$\begin{aligned} -H^0_{yx} &= -(D_{\omega_a}\mathcal{M})\mathcal{M}^{-1} + H^0_{yy} v_\theta \mathcal{M}^{-1} + f_{a,\theta}\mathcal{M}^{-1} \ , \\ \mathcal{M}^T H^0_{xx} &= -D_{\omega_a} v_\theta^T - v_\theta^T H^0_{yx} + g^T_{a,\theta} \ , \end{aligned} \tag{2.3.20}$$

and that, by definitions of z, w in (2.3.15) and $\mathcal{T}$ in (2.3.13), one has

$$\begin{aligned} \mathcal{M} D_{\omega_a}(\mathcal{M}^{-1} z) &= H^0_{yy}\mathcal{M}^{-T}(c + D^{-1}_{\omega_a} b_a) - f_a \ , \\ \mathcal{M}^T w &= v_\theta^T z + c + D^{-1}_{\omega_a} b_a \ . \end{aligned} \tag{2.3.21}$$

Then

$$\begin{aligned} & \omega_a + D_{\omega_a} u + D_{\omega_a} z - H_y(\theta + u + z, v + w) \\ & \overset{(i)}{=} \omega_a + D_{\omega_a} u + D_{\omega_a} z - H^0_y - H^0_{yx} z - H^0_{yy} w + Q_1 \\ & \overset{(ii)}{=} D_{\omega_a} z - H^0_{yx} z - H^0_{yy} w + f_a + Q_1 \\ & \overset{(iii)}{=} \mathcal{M}\ D_{\omega_a}(\mathcal{M}^{-1} z) + H^0_{yy} v_\theta\ \mathcal{M}^{-1} z - (H^0_{yy}\mathcal{M}^{-T})(\mathcal{M}^T w) + f_a + Q_1 + Q_2 \\ & \overset{(iv)}{=} H^0_{yy}\mathcal{M}^{-T}(\mathcal{M}^T v_\theta - v_\theta^T \mathcal{M})\mathcal{M}^{-1} z + Q_1 + Q_2 \\ & \overset{(v)}{=} Q_1 + Q_2 + Q_3 \\ & =: f' \ , \end{aligned} \tag{2.3.22}$$

where: (i) holds by definition of Q_1; (ii) holds by the first of (2.3.8); (iii) holds by the first in (2.3.20) and observing that

$$D_{\omega_a} z - (D_{\omega_a}\mathcal{M})\mathcal{M}^{-1} z = \mathcal{M} D_{\omega_a}(\mathcal{M}^{-1} z) \ ;$$

(iv) holds because of (2.3.21); (v) follows from the definition of $\mathcal{B}$ in (2.3.11), (2.3.12) and the definition of Q_3.

Notice that equating the third and the sixth line in (2.3.22) we have

$$D_{\omega_a} z - H^0_{yx} z - H^0_{yy} w + f_a = Q_2 + Q_3 \ . \tag{2.3.23}$$

Next,

$$\begin{aligned}
& D_{\omega_a} v + D_{\omega_a} w + H_x(\theta + u + z, v + w) \\
\overset{(i)}{=} \ & D_{\omega_a} v + D_{\omega_a} w + H^0_x + H^0_{xx} z + H^0_{xy} w + Q_4 \\
\overset{(ii)}{=} \ & \mathcal{M}^{-T} \mathcal{M}^T \ (D_{\omega_a} w + H^0_{xx} z + H^0_{xy} w + g_a) + Q_4 \\
\overset{(iii)}{=} \ & \mathcal{M}^{-T} \ \Big[D_{\omega_a}(\mathcal{M}^T w) - (D_{\omega_a} \mathcal{M}^T) w - (D_{\omega_a} v_\theta^T) z - v_\theta^T H^0_{yx} z \\
& + \mathcal{M}^T H^0_{xy} w + \mathcal{M}^T g_a \Big] + Q_4 + Q_5 \\
\overset{(iv)}{=} \ & \mathcal{M}^{-T} \ \Big[v_\theta^T D_{\omega_a} z + v_\theta^T f_a \ - (D_{\omega_a} \mathcal{M}^T) w - v_\theta^T H^0_{yx} z + \mathcal{M}^T H^0_{xy} w \Big] + Q_4 + Q_5 \\
\overset{(v)}{=} \ & \mathcal{M}^{-T} v_\theta^T \ \Big[D_{\omega_a} z + f_a - H^0_{yx} z - H^0_{yy} w \Big] + Q_4 + Q_5 + Q_6 \\
\overset{(vi)}{=} \ & \mathcal{M}^{-T} \ v_\theta^T (Q_2 + Q_3) + Q_4 + Q_5 + Q_6 \\
=: \ & Q_4 + Q_5 + Q_6 + Q_7 =: g' \ ,
\end{aligned}$$

where: (i) holds by definition of Q_4; (ii) holds by the second of (2.3.8); (iii) holds by the second in (2.3.20); (iv) holds because the definition of w implies that

$$D_{\omega_a}(\mathcal{M}^T w) = D_{\omega_a}(v_\theta^T z) + b_a \ ;$$

(v) follows from the transposed of the first equation in (2.3.10); finally, (vi) follows from (2.3.23). □

We turn, now, to discuss briefly system (2.3.14). We start with an elementary result, which allows to find a root α_0 of a (small and non-degenerate) function $F(\alpha)$. If α is a vector, $|\alpha|$ denotes the standard Euclidean norm, while if A is a matrix, $|A|$ denotes the standard norm $\sup_{|\alpha|=1} |A\alpha|$.

LEMMA 2.3.7. *Let* $\Omega := \{c \in \mathbb{R}^d : \ |c| \le \rho_1\} \times \{\hat{a} \in \mathbb{R} : \ |\hat{a}| \le \rho_2\}$ *for some* ρ_1, $\rho_2 > 0$; *let* $\alpha := (c, \hat{a}) \to F(\alpha) \in C^1(\Omega, \mathbb{R}^{d+1})$ *such that there exists* $F'(0)^{-1} =: \mathcal{B}$ *and such that*

$$\begin{aligned}
& \sup_\Omega |\alpha - \mathcal{B} F(\alpha)| \ \le \min\{\rho_1, \rho_2\} \ , \\
& \sup_\Omega |I - \mathcal{B} \ F'(\alpha)| < 1 \ .
\end{aligned} \tag{2.3.24}$$

Then, there exists a unique $\alpha_0 \in \Omega$ *such that* $F(\alpha_0) = 0$. *Moreover, one has*

$$\sup_{\alpha \in \Omega} |(F'(\alpha))^{-1}| \le \ |\mathcal{B}| \ \Big(1 - \sup_\Omega |I - \mathcal{B} F'|\Big)^{-1} \tag{2.3.25}$$

and

$$|\alpha_0| \le \ |\mathcal{B}| \ \Big(1 - \sup_\Omega |I - \mathcal{B} F'|\Big)^{-1} \ |F(0)| \ . \tag{2.3.26}$$

PROOF. Conditions (2.3.24) guarantee that $\alpha \in \Omega \to \phi(\alpha) := \alpha - \mathcal{B} F(\alpha)$ is a contraction on Ω: in particular, if $\phi := (\phi_1, \phi_2) \in \mathbb{R}^d \times \mathbb{R}$ and $|\phi| = \sqrt{|\phi_1|^2 + |\phi_2|^2}$, then the first equation in (2.3.24) yields $|\phi_1| \le \rho_1$ and $|\phi_2| \le \rho_2$, so that $\phi : \Omega \to \Omega$.

Thus, there exists a unique $\alpha_0 \in \Omega$ such that $\phi(\alpha_0) = \alpha_0$, namely $F(\alpha_0) = 0$. Furthermore,

$$|F'(\alpha)^{-1}| = |(I - (I - \mathcal{B}F'))^{-1}\, \mathcal{B}| \le |\mathcal{B}|\, (1 - |I - \mathcal{B}F'|)^{-1} ,$$

from which (2.3.25) follows. Inequality (2.3.26) follows from the identity

$$-F(0) = F(\alpha_0) - F(0) = \Big(\int_0^1 F'(t\alpha_0)dt\Big)\ \alpha_0 ,$$

which may be rewritten as

$$\alpha_0 = -\Big(I - \int_0^1 [I - \mathcal{B}F'(t\alpha_0)]dt\Big)^{-1} \mathcal{B}\, F(0) .$$

NOTATIONAL REMARK 2.3.8. For the purpose of the following discussion and the associated estimates, it will be useful to introduce a suitable (fixed) normalization parameter ρ having the physical dimension of the momenta y, and to rescale a (the number appearing in the definition of ω_a) by such parameter ρ, letting

$$\hat{a} := \rho a .$$

Define $F := (F_1, F_2) \in \mathbb{R}^d \times \mathbb{R}$ as

$$\begin{aligned} \text{(2.3.27)}\quad F_1(c,\hat{a}) &:= \Big(\langle \mathcal{T}\rangle c - \langle \mathcal{M}^{-1} f_a\rangle + \langle \mathcal{T} D_{\omega_a} b_a\rangle\Big)\Big|_{a=\frac{\hat{a}}{\rho}} \\ F_2(c,\hat{a}) &:= \frac{1}{\rho}\Big(H\left(0, v(0) + \mathcal{M}^{-T}(0)[c + D_{\omega_a}^{-1} b_a(0)]\right) - E\Big)\Big|_{a=\frac{\hat{a}}{\rho}} \\ &=: \frac{1}{\rho}\Big(H(0, v'(0; c, \hat{a})) - E\Big) , \end{aligned}$$

where

$$v'(0; c, \hat{a}) := v(0) + \mathcal{M}^{-T}(0)\Big[c + D_{\omega_a}^{-1} b_a(0)\Big]\Big|_{a=\frac{\hat{a}}{\rho}} .$$

Then, system (2.3.14) is equivalent to

$$F(\alpha) = 0 , \qquad \alpha := (c, \hat{a}) .$$

It is convenient to have "explicit" formulae concerning F.

LEMMA 2.3.9. *Let F be as in* (2.3.27) *(same notations as in Proposition 2.3.5). Then $F(0) = (F_1(0), F_2(0)$ with*

$$\begin{aligned} \text{(2.3.28)}\quad F_1(0) &= -\langle \mathcal{M}^{-1} f\rangle + \langle \mathcal{T} D^{-1} b\rangle , \\ F_2(0) &= \frac{1}{\rho}\Big[H(0, v(0)) - E + H_y(0, \tilde{v}_0)\mathcal{M}^{-T}(0) D^{-1} b(0)\Big] , \end{aligned}$$

where, for a suitable $t \in (0,1)$, we have set

$$\tilde{v}_0 := v(0) + t\ \mathcal{M}^{-T}(0) D^{-1} b(0) .$$

Let v_0' be short for

$$v_0' := v'(0; 0, 0) = v(0) + \mathcal{M}^{-T}(0) D^{-1} b(0) ,$$

and define the vector χ *and the matrix* $\mathcal{N}$ *as follows*

$$\chi(\theta) := \frac{1}{\rho}\mathcal{M}^{-1}H_y(\theta+u(\theta),v(\theta)),$$

$$\mathcal{N} := \begin{pmatrix} 0 & \mathcal{N}_{12} \\ \mathcal{N}_{21} & \mathcal{N}_{22} \end{pmatrix},$$

where

$$\begin{aligned} \mathcal{N}_{12} &:= \frac{-\langle \mathcal{M}^{-1}f\rangle - \langle \mathcal{T}D^{-1}(\partial_\theta H^0)\rangle}{\rho} \\ \mathcal{N}_{21} &:= \frac{\Big(H_y(0,v_0') - H_y(0,v(0))\Big)\mathcal{M}^{-T}(0)}{\rho} \\ \mathcal{N}_{22} &:= \frac{H_y(0,v_0')\cdot \mathcal{M}^{-T}(0)D^{-1}(\partial_\theta H^0)(0)}{\rho^2}. \end{aligned}$$

Then,

$$F'(0) = \begin{pmatrix} \langle \mathcal{T}\rangle & -\langle\chi\rangle \\ \chi(0)^T & 0 \end{pmatrix} + \mathcal{N}. \tag{2.3.29}$$

Define, now, $\widetilde{F}$ *as the second-order remainder of* F*:*

$$F(\alpha) = F(0) + F'(0)\alpha + \widetilde{F}(\alpha). \tag{2.3.30}$$

Then[2.7]*,*

$$\begin{aligned} \widetilde{F}(\alpha) &= \Big(-\frac{a^2}{1+a}\langle \mathcal{T}D^{-1}(\partial_\theta H^0)\rangle\ ,\ \frac{1}{\rho}\Big[\frac{1}{2}H_{yy}(0,\tilde v_0)\delta_{v'}\cdot\delta_{v'} \\ &\quad - \frac{a^2}{1+a}H_y(0,v'(0;0))\cdot\mathcal{M}^{-T}(0)D^{-1}(\partial_\theta H^0)(0)\Big]\Big), \end{aligned} \tag{2.3.31}$$

where, for some $0 < t < 1$,

$$\tilde v_0 := v'(0;0) + t(v'(0;\alpha) - v'(0;0)), \qquad \delta_{v'} := v'(0;\alpha) - v'(0;0).$$

The Jacobian of $\widetilde{F}$ *has the form*

$$\widetilde{F}_\alpha(\alpha) = \frac{1}{\rho}\begin{pmatrix} 0 & \widetilde{F}_{12} \\ \delta_H\mathcal{M}^{-T}(0) & \widetilde{F}_{22} \end{pmatrix}, \tag{2.3.32}$$

where

$$\begin{aligned} \delta_H &:= H_y(0,v'(0;\alpha)) - H_y(0,v'(0;0)) = H_{yy}(0,\tilde v_0')\delta_{v'} \\ \widetilde{F}_{12} &:= -a\,\frac{2+a}{(1+a)^2}\langle \mathcal{T}D^{-1}(\partial_\theta H^0)\rangle \\ \widetilde{F}_{22} &:= \Big(\frac{\delta_H}{\rho} - \frac{a}{\rho}\,\frac{2+a}{(1+a)^2}H_y(0,v'(0;\alpha))\Big)\cdot\mathcal{M}^{-T}(0)D^{-1}(\partial_\theta H^0)(0) \end{aligned}$$

[2.7] Recall that $a = \hat a/\rho$.

and

$$\tilde{v}_0' := v'(0;0) + t'(v'(0;\alpha) - v'(0;0))$$

for some $0 < t' < 1$. *Finally,*

$$\partial_\theta H(\theta + u, v) = \partial_\theta D^{-1}(H_x^0 \cdot f + H_y^0 \cdot g) \ . \tag{2.3.33}$$

REMARK 2.3.10. (i) In view of (2.3.29) (the matrix $\mathcal{N}$ has to be thought small, see next point of this remark), one sees that the non-degeneracy condition to be required in order to solve (2.3.14) is

$$\det \mathcal{A} \neq 0 \ , \tag{2.3.34}$$

where

$$\mathcal{A} := \begin{pmatrix} \langle \mathcal{T} \rangle & -\langle \chi \rangle \\ \chi(0)^T & 0 \end{pmatrix}, \quad \mathcal{T} := \mathcal{M}^{-1} H_{yy}^0 \mathcal{M}^{-T},$$

$$\mathcal{M} := I + u_\theta, \quad \chi := \frac{1}{\rho} \mathcal{M}^{-1} H_y^0. \tag{2.3.35}$$

The non-degeneracy condition (2.3.34), in the nearly-integrable case

$$H = H_0(y) + \varepsilon H_1(x,y) \ ,$$

will be recognized to be, in the limiting case $\varepsilon = 0$ (taking as approximate torus the unperturbed torus $u \equiv 0$, $v \equiv y$), *Arnold's iso-energetic non-degeneracy condition*

$$\det \begin{pmatrix} H_0'' & H_0' \\ H_0' & 0 \end{pmatrix} \neq 0 \ ; \tag{2.3.36}$$

compare next item.

(ii) Let us briefly discuss, here, the geometrical meaning of Arnold's iso-energetic non-degeneracy condition.

The standard KAM non-degeneracy condition for $H_0 : V \subset \mathbb{R}^d \to \mathbb{R}$, at a point $y_0 \in V$, can be formulated by saying that the frequency map

$$\omega : y \in V \to \omega(y) := \frac{\partial H_0}{\partial y}(y) \in \mathbb{R}^n \ ,$$

is a local smooth diffeomorphism near y_0, or equivalently ("infinitesimal version"), that

$$\partial_y \omega \ (TV_{y_0}) = T\mathbb{R}^d_{y_0} \ , \qquad (y_0 \in V) \ ,$$

where $TV_{y_0} = \mathbb{R}^d$ denotes the tangent space of V at y_0. The (infinitesimal version of the) iso-energetic non-degeneracy condition for H_0 at a point y_0 on the energy level $H_0^{-1}(E)$, can be formulated by saying that

$$\partial_y \omega \ (TS_{E,y_0}) \oplus N_{y_0} = T\mathbb{R}^d_{y_0} \ , \qquad (y_0 \in S_E) \ , \tag{2.3.37}$$

where TS_{E,y_0} denotes the tangent space of S_E at y_0 and $N_{y_0} := \{\lambda \omega(y_0) | \ \lambda \in \mathbb{R}\}$ denotes the normal space at y_0 to TS_{E,y_0}. Notice that (2.3.37) requires implicitly that $H_0^{-1}(E)$ is smooth ipersurface in a neighborhood of y_0.

Clearly, condition (2.3.37) is equivalent to require (2.3.36) at the point $y_0 \in S_E$. Notice, also, that the matrix in (2.3.36) is the Jacobian at $(y,a) = (y_0, 0)$ of the "frequency-energy map"

$$\phi : (y,a) \to (\omega, E) := \Big((1+a)\omega(y), H_0(y) \Big) \ .$$

Another version of Arnold's iso-energetic non-degeneracy condition (corresponding to ask that $\partial_y \omega$ is a local diffeomorphism in the non-fixed-energy case) is to require that the map

$$\Omega_E : y \in S_E \to \Omega_E(y) := \pi \circ \omega(y) \ ,$$

where π is the canonical projection of $\mathbb{R}^d$ onto the real $d-1$ dimensional projective space $\mathbb{P}^{d-1}$, is a local smooth diffeomorphism near $y_0 \in S_E$, i.e., that the rank of the differential $\Omega_{E,*}$ of the map Ω_E at $y_0 \in S_E$ is maximal $(d-1)$.

Let us show, in fact, that *the vanishing of the determinant in* (2.3.36) *at a regular point* $y_0 \in S_E$ *is equivalent to have* $\mathrm{rank}(\Omega_{E,*y_0}) < d-1$. We prove this claim using local coordinates. Since y_0 is a regular point of $S_E = H_0^{-1}(E)$, there exists $1 \le i \le d$ such that $\partial_{y_i} H_0(y_0) = \omega_i(y_0) \ne 0$. Let us assume (without loss of generality) that $i = d$. Then, as local coordinates on S_E around y_0 we can take $\hat{y} := (y_1, ..., y_{d-1})$ since, by the Implicit Function Theorem, there exists a (unique) smooth function g from a neighborhood of $\hat{y}_0$ into a neighborhood of y_{0d} such that $H_0(\hat{y}, g(\hat{y})) \equiv E$. As local coordinates in a neighborhood of $\pi(\omega_0) := \pi(\omega(y_0))$ we can take

$$\hat{u} := (u_1, ..., u_{d-1}) \in U_0 \to \pi\Big((\hat{u}, 1)\Big) \ ,$$

where U_0 is a neighborhood of $\Big(\omega_{01}/\omega_{0d}, ..., \omega_{0(d-1)}/\omega_{0d}\Big)$. In such coordinates, as it is easy to check, the $(d-1) \times (d-1)$ matrix associated to the differential $\Omega_{E,*}$ is given by

$$\hat{\Omega}_{*ij} = \frac{1}{\omega_d}\left(\frac{\partial \omega_i}{\partial y_j} + \frac{\omega_i \omega_j}{\omega_d^2}\frac{\partial \omega_d}{\partial y_d} - \frac{\omega_j}{\omega_d}\frac{\partial \omega_d}{\partial y_i} - \frac{\omega_i}{\omega_d}\frac{\partial \omega_d}{\partial y_j}\right) \ , \qquad (1 \le i, j \le d-1)$$

(everything is evaluated at $y_0 \in S_E$). On the other hand, the vanishing of the determinant in (2.3.36) at y_0 is equivalent to say that there exists $\xi \in \mathbb{R}^d \backslash \{0\}$ and $\lambda \in \mathbb{R}$ such

$$\frac{\partial \omega}{\partial y}(y_0)\, \xi = \lambda \omega_0 \ , \qquad \omega_0 \cdot \xi = 0 \ , \tag{2.3.38}$$

which is equivalent to

$$\xi_d = -\frac{1}{\omega_d}\sum_{j=1}^{d-1} \omega_j \xi_j \ , \qquad \lambda = \frac{1}{\omega_d}\sum_{j=1}^{d-1}\frac{\partial \omega_d}{\partial y_j}\xi_j - \frac{1}{\omega_d^2}\frac{\partial \omega_d}{\partial y_d}\sum_{j=1}^{d-1}\omega_j \xi_j \ ,$$

$$\sum_{j=1}^{d-1}\frac{\partial \omega_i}{\partial y_j}\xi_j + \frac{\partial \omega_i}{\partial y_d}\xi_d = \lambda \omega_i \ , \quad 1 \le i \le d-1 \ , \tag{2.3.39}$$

(everything is evaluated at $y_0 \in S_E$). Therefore, if the determinant in (2.3.36) vanishes, then taking $\hat{\xi} := (\xi_1, ..., \xi_{d-1})$ with ξ as in (2.3.38), from (2.3.39) it follows that $\hat{\Omega}_* \hat{\xi} = 0$; vice-versa, if $\hat{\xi} \in \mathbb{R}^{d-1}\backslash\{0\}$ is such that $\hat{\Omega}_* \hat{\xi} = 0$, then, *defining* ξ_d and λ as in the first line of (2.3.39), one sees that (2.3.38) holds (with $\xi \in \mathbb{R}^d \backslash \{0\}$). □

(iii) Recall the philosophy of Remark 2.3.6, point (iii); in particular imagine to replace the error functions f and g, respectively, by ηf and ηg with some small parameter η. *Assume also the non-degeneracy condition* (2.3.34) *and let* $|c| \le \bar{\eta}$, $|\hat{a}| \le \bar{\eta}$ *with*

$$\eta \ll \bar{\eta} \ll 1 \ . \tag{2.3.40}$$

Then (see also (2.3.33)),

$$b := b_0 = O(\eta) \ , \qquad \partial_\theta H(\theta + u, v) = O(\eta) \ , \tag{2.3.41}$$

and, by (2.3.28), (2.3.31), (2.3.32) and (2.3.29), one finds easily

$$F(0) = O(\eta) \ , \quad \tilde{F} = O(\bar{\eta}^2) \ , \quad \tilde{F}_\alpha = O(\bar{\eta}) \ , \quad \mathcal{N} = O(\bar{\eta}) \ . \tag{2.3.42}$$

Thus, $F'(0)$ is invertible and, denoting by $\mathcal{B}$ its inverse, we find

$$\mathcal{B} := \Big(F'(0)\Big)^{-1} = \mathcal{A}^{-1} + O(\bar{\eta}) \ . \tag{2.3.43}$$

Furthermore, recalling the definition of $\tilde{F}$,

$$
\begin{aligned}
|\alpha - \mathcal{B}F(\alpha)| &= |\mathcal{B}F(0) + \mathcal{B}\widetilde{F}(\alpha)| \le |\mathcal{B}| \left(|F(0)| + |\widetilde{F}(\alpha)|\right) \\
|I - \mathcal{B}F'(\alpha)| &= |\mathcal{B}\widetilde{F}_\alpha| \le |\mathcal{B}|\ |\widetilde{F}_\alpha(\alpha)|\ . \qquad (2.3.44)
\end{aligned}
$$

Thus, we see that (2.3.40)÷(2.3.44) allow to apply Lemma 2.3.7 so as to obtain a root $\alpha_0 = (c_0, \hat{a}_0)$ of F satisfying (2.3.26), i.e.,

$$(c_0, \hat{a}_0) = O(\eta)\ .$$

Notice in particular that such c_0 and $\hat{a}_0$ verify (2.3.18), which was assumed in Remark 2.3.6, point (iii).

PROOF. (of Lemma 2.3.9) From the definition (2.3.27)

$$F(0) = \left(-\langle \mathcal{M}^{-1} f\rangle + \langle \mathcal{T} D^{-1} b\rangle, \frac{H\Big(0, v(0) + \mathcal{M}^{-T}(0) D^{-1} b(0)\Big) - E}{\rho} \right),$$

and equality (2.3.28) follows immediately from Lagrange formula.

We compute now the linear part of F. Recalling (2.3.7) and the definitions of b and b_a (i.e., (2.3.3) and (2.3.11)), one finds that

$$b_a = (1+a)b + a(\partial_\theta H^0)$$

and therefore

$$
\begin{aligned}
D_{\omega_a}^{-1} b_a &= D^{-1} b + a D_{\omega_a}^{-1}(\partial_\theta H^0) \\
&= D^{-1} b + \frac{a}{1+a} D^{-1}(\partial_\theta H^0) \\
&= D^{-1} b + a D^{-1}(\partial_\theta H^0) - \frac{a^2}{1+a} D^{-1}(\partial_\theta H^0)\ .
\end{aligned}
$$

Next, we remark that

$$
\begin{aligned}
&\partial_a f_a = f + H^0_y\ , \\
&\partial_a g_a = g - H^0_x\ , \\
&\partial_a b_a = b + \partial_\theta H^0\ , \\
&\partial_a (D_{\omega_a}^{-1} b_a) = D_{\omega_a}^{-1}(\partial_\theta H^0) - \frac{a}{1+a} D_{\omega_a}^{-1}(\partial_\theta H^0) \\
&\qquad = \frac{1}{1+a^2} D^{-1}(\partial_\theta H^0)\ , \\
&\partial_a \langle \mathcal{M}^{-1} f_a - \mathcal{T} D_{\omega_a}^{-1} b_a\rangle = \langle \mathcal{M}^{-1} f\rangle + \langle \mathcal{M}^{-1} H^0_y\rangle - \frac{1}{(1+a)^2}\langle \mathcal{T} D^{-1}(\partial_\theta H^0)\rangle\ .
\end{aligned}
$$

Therefore, since $\partial_{\hat{a}} = \frac{1}{\rho}\partial_a$, we find

$$F'(\alpha) = \begin{pmatrix} \langle \mathcal{T}\rangle & F'_{12} \\ \\ \dfrac{H_y(0, v'(0; c, \hat{a}))\mathcal{M}^{-T}(0)}{\rho} & F'_{22} \end{pmatrix},$$

where

$$\begin{aligned} v'(0;c,\hat{a}) &= v(0) + \mathcal{M}^{-T}(0)D^{-1}b + \mathcal{M}^{-T}(0)\left(c + \frac{a}{1+a}D^{-1}(\partial_\theta H^0)\right) \\ &= v'(0;0,0) + \mathcal{M}^{-T}(0)\left(c + aD^{-1}(\partial_\theta H^0) - \frac{a^2}{1+a}D^{-1}(\partial_\theta H^0)\right) \end{aligned}$$

$$\begin{aligned} F'_{12} &:= -\frac{\langle \mathcal{M}^{-1}f\rangle + \langle \mathcal{M}^{-1}H_y^0\rangle - \dfrac{\langle \mathcal{T}D^{-1}(\partial_\theta H^0)\rangle}{(1+a)^2}}{\rho} \\ F'_{22} &:= \frac{H_y(0,v'(0;c,\hat{a}))\cdot \mathcal{M}^{-T}(0)(D^{-1}\partial_\theta H^0)(0)}{\rho^2(1+a)^2} . \end{aligned}$$

Setting $(c,\hat{a}) = (0,0)$, one obtains immediately (2.3.29).

Subtracting the linear part $F(0) + F'(0)\alpha$ to F, one obtains the expression for $\tilde{F}$ given in (2.3.31).

Notice that from the definition of $\tilde{F}$, (2.3.30), it follows that $\tilde{F}_\alpha(\alpha) = F'(\alpha) - F'(0)$ and a straightforward computation yields (2.3.32).

Finally, (2.3.33) is obtained by applying $\partial_\theta D^{-1}(\cdot)$ in the following relation:

$$\begin{aligned} DH(\theta+u,v) &= H_x^0\cdot(\omega + Du) + H_y^0\cdot Dv \\ &= H_x^0\cdot(H_y^0 + f) + H_y^0\cdot(-H_x^0 + g) \\ &= H_x^0\cdot f + H_y^0\cdot g . \quad \blacksquare \end{aligned}$$

2.4. The KAM Map

We now define the KAM functional map as follows. Let $\omega \in \mathbb{R}^d$ be a (γ,τ)-Diophantine vector (see (2.1.3)) and let us consider an approximate KAM torus (u,v,ω) as in (2.3.1), satisfying (2.3.34), i.e., following the notations in § 2.3, $u,v : \mathbb{T}^d \to \mathbb{R}^d$ are real-analytic function satisfying

(2.4.1) $$\det \mathcal{M}(\theta) \neq 0, \quad \forall\, \theta \in \mathbb{T}^d, \quad \mathcal{M}(\theta) := I + u_\theta;$$

(2.4.2) $$\det \mathcal{A} \neq 0, \quad \mathcal{A} := \begin{pmatrix} \langle \mathcal{T}\rangle & -\langle\chi\rangle \\ \chi(0)^T & 0 \end{pmatrix},$$

$$\mathcal{T} := \mathcal{M}^{-1}H_{yy}^0\mathcal{M}^{-T}, \quad \chi := \frac{1}{\rho}\mathcal{M}^{-1}H_y^0;$$

$$u(0) = 0 ;$$

where $\rho > 0$ is a prefixed weight and the suffix ^{0} means, here, that the argument of the function is given by $\big(\theta + u(\theta), v(\theta)\big)$. Let f, g and h be defined by the following "approximate torus equations":

$$\begin{aligned} \omega + D_\omega u - H_y(\theta+u,v) &=: f , \\ D_\omega v + H_x(\theta+u,v) &=: g , \\ u(0) &= 0 , \end{aligned}$$

(2.4.3) $$H(0,v(0)) - E =: h .$$

The *functional KAM map* is defined to be the map

(2.4.4) $$\mathcal{K} : (u,v,\omega) \mapsto (u',v',\omega') := \big(u+z, v+w, (1+a)\omega\big) ,$$

where z, w and a are as in Proposition 2.3.5.

In the next sections we shall equip the functional KAM map with (careful) estimates.

2.5. Technical Tools

In this section we fix some notation and review a few technical facts, which are needed in the estimates performed in the rest of this paper.

2.5.1. Norms on numbers, matrices and tensors. If $a \in \mathbb{C}^d$, we denote by $|\cdot|$ the standard 2-norm

$$|a| := \sqrt{\sum_{i=1}^{d} |a_i|^2} \; .$$

If $A = (a_{ij}) \in \mathrm{Mat}(n \times m)$, then

$$|A| := \sup_{a \in \mathbb{C}^n : |a|=1} |Aa| \; . \tag{2.5.1}$$

Notice that

$$|A| \le \sqrt{\sum |a_{ij}|^2} \; . \tag{2.5.2}$$

In particular, if $a, b \in \mathbb{C}^n$ and if we denote by $a \otimes b$ the matrix with entries $(a \otimes b)_{ij} = a_i b_j$, then

$$|a \otimes b| \le |a||b| \; .$$

In the coming sections we shall also use the following elementary bounds on the (Euclidean) norm of a matrix $A \in \mathrm{Mat}((d+1) \times (d+1))$: write A as

$$A = \begin{pmatrix} A_{11} & a_{12} \\ a_{21} & a_{22} \end{pmatrix}$$

with $A_{11} \in \mathrm{Mat}(d \times d)$, $a_{12}, a_{21} \in \mathbb{C}^d$, $a_{22} \in \mathbb{C}$; then[2.8]

$$|A| \le \sqrt{|A_{11}|^2 + |a_{12}|^2 + |a_{21}|^2 + |a_{22}|^2} \; , \tag{2.5.3}$$

$$|A| \le \sqrt{\max\{|A_{11}|^2 + |a_{21}|^2, |a_{12}|^2 + |a_{22}|^2\} + |A_{11}^T a_{12}| + |a_{21}| \; |a_{22}|} \; . \tag{2.5.4}$$

Finally, if T is a 3-tensor, i.e., a linear map from $\mathbb{C}^d$ into $\mathrm{Mat}(n \times m)$, then

$$|T| := \sup_{a \in \mathbb{C}^d : |a|=1} |Ta| = \sup_{a \in \mathbb{C}^d : |a|=1} \; \sup_{b \in \mathbb{C}^m : |b|=1} |(Ta)b| \; . \tag{2.5.5}$$

[2.8] The two inequalities (2.5.3) and (2.5.4) are independent, as one checks immediately on the two matrices $\begin{pmatrix} 1 & 0 \\ 0 & 1 \end{pmatrix}$ and $\begin{pmatrix} 1 & 2 \\ 1 & 1 \end{pmatrix}$: in the first case (2.5.4) is better, in the second case (2.5.3) is better. Let us check here (2.5.4) ((2.5.3) is left to the reader). Let $x \in \mathbb{C}^d$ and $y \in \mathbb{C}$ s.t. $|x|^2 + |y|^2 = 1$ then:

$$\begin{aligned} \left| A \begin{pmatrix} x \\ y \end{pmatrix} \right|^2 &\le |A_{11}x|^2 + |a_{12}|^2|y|^2 + 2|y||A_{11}^T a_{12} \cdot x| + |a_{21} \cdot x|^2 + |a_{22}| \, |y|^2 + 2|a_{21} \cdot x||a_{22}| \, |y| \\ &\le (|A_{11}|^2 + |a_{21}|^2)|x|^2 + (|a_{12}|^2 + |a_{22}|^2)|y|^2 + (|A_{11}^T a_{12}| + |a_{22}||a_{21}|)2|x||y| \\ &\le \max\{|A_{11}|^2 + |a_{21}|^2, |a_{12}|^2 + |a_{22}|^2\} + (|A_{11}^T a_{12}| + |a_{22}||a_{21}|). \end{aligned}$$

REMARK 2.5.1. (i) The space $\mathbb{C}^d$, thought of as a vector space over $\mathbb{C}$, will always be endowed with the standard orthonormal basis $(1,0,...,0)$, $(0,1,0,...)$,..., $(0,...,0,1)$. Therefore, we shall identify linear maps from $\mathbb{C}^m$ into $\mathbb{C}^n$ with $\mathrm{Mat}_{\mathbb{C}}(n\times m)$ and 3-tensors with "three-index-objects". In particular if T is a 3-tensor, Ta denotes the $(n\times m)$ matrix with entries

$$(Ta)_{ij}=\sum_{k=1}^{d} T_{ijk}a_k \ .$$

(ii) Notice that if T is a matrix or a tensor with non-negative entries, then $\mathbb{C}$ in (2.5.1) and (2.5.5) can be replaced by $\mathbb{R}_+ := [0,\infty)$.

(iii) To unify notations, we let $X_1 = X_1(d)$, $X_2 = X_2(n,m)$ and $X_3 = X_3(n,m,d)$ denote, respectively, $\mathbb{C}^d$, $\mathrm{Mat}_{\mathbb{C}}(n\times m)$ and the space of linear maps from $\mathbb{C}^d$ into $\mathrm{Mat}_{\mathbb{C}}(n\times m)$ (i.e., the space of 3-tensors).

2.5.2. Norms on analytic functions. Let $D_r^m(y_0)$ and $\mathbb{T}_\xi^m$ denote the complex sets given by

$$D_r^m(y_0) := \Big\{ y\in\mathbb{C}^m \ : \ |y_i - y_{0i}|\le r \ , \ \forall\, i\Big\} \ ,$$
$$\mathbb{T}_\xi^m := \Big\{ y\in\mathbb{C}^m \ : \ |\mathrm{Im}\, y|\le \xi \ , \quad \mathrm{Re}\, y_i \mod 2\pi\Big\} \ .$$

If f is an analytic function, $f:\mathbb{T}_\xi^d\to\mathbb{C}$, with Fourier expansion

$$f(x)=\sum_{n\in\mathbb{Z}^d} f_n\ \exp(in\cdot x) \ , \qquad f_n := \int_{\mathbb{T}^d} f(\theta)\exp(-in\cdot\theta)\ \frac{d\theta}{(2\pi)^d} \ ,$$

we set

$$\|f\|_\xi := \sum_{n\in\mathbb{Z}^d} |f_n|\exp(|n|\xi) \ .$$

If f is an analytic function, $f:\mathbb{T}_\xi^d\times D_r^m(y_0)\to\mathbb{C}$, with Taylor-Fourier expansion[2.9]

$$f(x,y)=\sum_{\substack{n\in\mathbb{Z}^d\\ k\in\mathbb{N}^d}} f_{nk}\ (y-y_0)^k\ \exp(in\cdot x) \ ,$$

we set

$$\|f\|_{\xi,r} := \sum_n |f_n|_r \exp(|n|\xi) := \sum_{n,k} |f_{nk}| r^{|k|_1}\exp(|n|\xi) \ ,$$

where

$$|k|_1 := \sum_{j=1}^{d} |k_j| \ .$$

Notice that

$$\sup_{\mathbb{T}_\xi^d\times D_r^m(y_0)} |f|\le \|f\|_{\xi,r} \ .$$

The above definitions generalize immediately to the case in which f takes value into spaces of vectors, matrices or tensors.

If f is a X_q-valued, real-analytic function on $\mathbb{T}_\xi^d$, we let $N_\xi(f)$ denote the element (with non-negative entries) in X_q defined, for $q=1,2,3$, respectively, as

$$N_\xi(f)_i = \|f_i\|_\xi \ , \quad N_\xi(f)_{ij} = \|f_{ij}\|_\xi \ , \quad N_\xi(f)_{ijk} = \|f_{ijk}\|_\xi \ ; \tag{2.5.6}$$

then, we define

$$\|f\|_\xi := |N_\xi(f)| \ . \tag{2.5.7}$$

[2.9] As standard, $(y-y_0)^k := (y_1-y_{01})^{k_1}\cdots(y_m-y_{0m})^{k_m}$.

For example, if f takes value in $X_2(n,m) = \mathrm{Mat}_{\mathbb{C}}(n \times m)$, it is[2.10]

$$\|f\|_\xi = \sup_{\substack{a\in\mathbb{R}^m_+ \\ |a|=1}} \sqrt{\sum_{i=1}^n \Big(\sum_{j=1}^m \|f_{ij}\|_\xi a_j\Big)^2} \ .$$

If f is real-analytic on $\mathbb{T}^d_\xi \times D^m_r(y_0)$ we simply replace $\|\cdot\|_\xi$ with $\|\cdot\|_{\xi,r}$ in (2.5.6) and (2.5.7).

Consistently, if $f = \sum_{k\in\mathbb{N}^d} f_k\ (y-y_0)^k$ is real-analytic on $D^m_r(y_0)$, we set

$$\|f\|_r := \sum_{k\in\mathbb{N}^d} |f_k| r^{|k|_1} \ .$$

Clearly, if f is a X_q-valued function on $\mathbb{T}^d_\xi \times D^m_r(y_0)$, a simple bound on its norm is given by

$$\|f\|_{\xi,r} \le \sqrt{\sum_{i_1,\dots,i_q} \|f_{i_1\cdots i_q}\|^2_{\xi,r}} \ . \tag{2.5.8}$$

2.5.3. Banach spaces of real-analytic functions. Let Ω be either $\mathbb{T}^d_\xi$ or $\mathbb{T}^d_\xi \times D^m_r(y_0)$ and denote by $\mathcal{R}(\Omega, X_q)$ the space of X_q-valued, real-analytic functions on Ω with finite norm $\|\cdot\|_\xi$ or $\|\cdot\|_{\xi,r}$. The space $\mathcal{R}(\Omega, X_q)$ is a *Banach space*, as one immediately checks.

2.5.4. Product of real-analytic functions. We first notice that $\mathcal{R}(\mathbb{T}^d_\xi, \mathbb{C})$ is a *Banach algebra*, that is

$$\|fg\|_\xi \le \|f\|_\xi\ \|g\|_\xi \ , \qquad \forall\ f,g \in \mathcal{R}(\mathbb{T}^d_\xi, \mathbb{C}) \ ; \tag{2.5.9}$$

in fact:

$$\begin{aligned}
\|fg\|_\xi &:= \sum_n |(fg)_n| \exp(|n|\xi) \\
&= \sum_n \Big|\sum_m f_m g_{n-m}\Big| \ \exp(|n|\xi) \\
&\le \sum_{n,m} \exp(|m|) \exp(|n-m|) |f_m|\ |g_{n-m}| \\
&= \|f\|_\xi \|g\|_\xi \ .
\end{aligned}$$

Relation (2.5.9) generalizes to the X_q-valued case, provided the product fg is well defined (as in the case when $f \in X_2(n,m)$ or $f \in X_3(\ell,n,m)$ and $g \in X_1(m)$). For example, let us check the inequality in (2.5.9) in the case $f \in \mathcal{R}(\mathbb{T}^d_\xi, \mathrm{Mat}_{\mathbb{C}}(n\times m))$, $g \in \mathcal{R}(\mathbb{T}^d_\xi, \mathbb{C}^m)$:

$$\begin{aligned}
\|fg\|_\xi &:= |N_\xi(fg)| \\
&= \sqrt{\sum_{i=1}^n \|\sum_{j=1}^m f_{ij} g_j\|^2_\xi} \\
&\le \sqrt{\sum_{i=1}^n (\sum_{j=1}^m \|f_{ij}\|_\xi \|g_j\|_\xi)^2} \\
&= |N_\xi(f)\ N_\xi(g)| \\
&\le |N_\xi(f)|\ |N_\xi(g)| \\
&= \|f\|_\xi\ \|g\|_\xi \ ,
\end{aligned}$$

(in the first inequality we used (2.5.9)).

[2.10] Recall point (ii) of Remark 2.5.1.

Analogous estimates hold in the case of product of analytic functions on $D_r^m(y_0)$: if $f, g \in \mathcal{R}(D_r^m(y_0), \mathbb{C})$, then[2.11]

$$\|fg\|_r \le \|f\|_r \ \|g\|_r \ .$$

To conclude this paragraph, we give some explicit examples of evaluations of norms:

$$\begin{aligned} &\|\exp(n\cdot x)\|_\xi \ , \ \|\sin(n\cdot x)\|_\xi \ , \ \|\cos(n\cdot x)\|_\xi \ = \ \exp(|n|\xi) \ , \\ &\|f(y)\sin(n\cdot x)\|_{\xi,r} = \|f\|_r \exp(|n|\xi) \ , \\ &\|f(y)\cos(n\cdot x)\|_{\xi,r} = \|f\|_r \exp(|n|\xi) \ , \end{aligned} \tag{2.5.10}$$

and, more in general,

$$\begin{aligned} &\Big\|\sum_n f_n(y)\exp(n\cdot x)\Big\|_{\xi,r} \ , \Big\|\sum_n f_n(y)\sin(n\cdot x)\Big\|_{\xi,r} \ , \Big\|\sum_n f_n(y)\cos(n\cdot x)\Big\|_{\xi,r} \\ &\qquad \le \sum_n \|f_n\|_r \exp(|n|\xi) \ . \end{aligned} \tag{2.5.11}$$

2.5.5. Composition of analytic functions.

Let $0 < \xi < \bar\xi$ and let $h \in \mathcal{R}(\mathbb{T}^n_{\bar\xi} \times D_r^m(y_0), X_q)$, $f \in \mathcal{R}(\mathbb{T}^d_\xi, \mathbb{C}^n)$, $g \in \mathcal{R}(\mathbb{T}^d_\xi, \mathbb{C}^m)$. Assume that

$$\|f\|_\xi \le \bar\xi - \xi \ , \qquad \|g_s - y_{0s}\|_\xi \le r \ , \quad (1 \le s \le m) \ , \tag{2.5.12}$$

and define

$$\phi : \theta \in \mathbb{T}^d_\xi \to \phi(\theta) := \Big(\theta + f(\theta), g(\theta)\Big) \in \mathbb{C}^{n+m} \ .$$

Then, $h\circ\phi \in \mathcal{R}(\mathbb{T}^d_\xi, X_q)$ and

$$\|h\circ\phi\|_\xi \le \|h\|_{\bar\xi,r} \ .$$

PROOF. We first consider the special case $X_q = X_1(1) = \mathbb{C}$. Using (repeatedly) the fact that $\mathcal{R}(\mathbb{T}^d_\xi, \mathbb{C})$ is a Banach algebra and using (in the final inequality) the assumptions (2.5.12), we find[2.12]

[2.11] In fact:

$$\begin{aligned} \|fg\|_r \ &:= \ \sum_k |(fg)_k| \ r^{|k|_1} = \sum_k \Big| \sum_{l+m=k} f_l r^{|l|_1} g_m r^{|m|_1} \Big| \\ &\le \ \sum_k \sum_{l+m=k} |f_l| r^{|l|_1} |g_m| r^{|m|_1} = \|f\|_r \ \|g\|_r \ . \end{aligned}$$

[2.12] The indices ℓ', ℓ, k and j run over, respectively, $\mathbb{Z}^d$, $\mathbb{Z}^d$, $\mathbb{N}^m$ and $\mathbb{N}$.

$$
\begin{aligned}
\|h\circ\phi\|_\xi &= \sum_{\ell'}\Big|\sum_{\ell,k,j}\frac{h_{\ell,k}}{j!}\,i^j\,\Big((\ell\cdot f)^j(g-y_0)^k\Big)_{\ell'-\ell}\Big|\ \exp(|\ell'|\xi)\\
&\le \sum_{\ell',\ell,k,j}\exp(|\ell|\xi)\frac{|h_{\ell,k}|}{j!}\ \Big|\Big((\ell\cdot f)^j(g-y_0)^k\Big)_{\ell'-\ell}\Big|\ \exp(|\ell'-\ell|\xi)\\
&= \sum_{\ell,k,j}\frac{|h_{\ell,k}|}{j!}\ \|(\ell\cdot f)^j(g-y_0)^k\|_\xi\exp(|\ell|\xi)\\
&\le \sum_{\ell,k,j}\frac{|h_{\ell,k}|}{j!}\ \|\ell\cdot f\|_\xi^j\prod_{s=1}^m\|g_s-y_{0s}\|_\xi^{k_s}\ \exp(|\ell|\xi)\\
&\le \sum_{\ell,k,j}\frac{|h_{\ell,k}|}{j!}\ \Big(\sum_{s'=1}^n|\ell_{s'}|\|f_{s'}\|_\xi\Big)^j\prod_{s=1}^m\|g_s-y_{0s}\|_\xi^{k_s}\ \exp(|\ell|\xi)\\
&\le \sum_{\ell,k,j}\frac{|h_{\ell,k}|}{j!}\ (|\ell|\ |N_\xi(f)|)^j\prod_{s=1}^m\|g_s-y_{0s}\|_\xi^{k_s}\ \exp(|\ell|\xi)\\
&= \sum_{\ell,k,j}\frac{|h_{\ell,k}|}{j!}\ (|\ell|\ \|f\|_\xi)^j\prod_{s=1}^m\|g_s-y_{0s}\|_\xi^{k_s}\ \exp(|\ell|\xi)\\
&\le \sum_{\ell,k,j}\frac{|h_{\ell,k}|}{j!}\ |\ell|^j\ (\bar\xi-\xi)^j\ r^{|k|_1}\exp(|\ell|\xi)\\
&= \|h\|_{\bar\xi,r}\ . \qquad (2.5.13)
\end{aligned}
$$

The case when h takes values in X_q follows easily; as an example let us check the case when h is matrix-valued, i.e., $X_q = X_2$:

$$
\begin{aligned}
\|h\circ\phi\|_\xi &= |N_\xi(h\circ\phi)| = \sup_{|c|=1}|N_\xi(h\circ\phi)c|\\
&= \sup_{|c|=1}\sqrt{\sum_i|\sum_j N_\xi(h\circ\phi)_{ij}c_j|^2}\\
&= \sup_{|c|=1}\sqrt{\sum_i|\sum_j \|h_{ij}\circ\phi\|_\xi c_j|^2}\\
&\le \sup_{|c|=1}\sqrt{\sum_i|\sum_j \|h_{ij}\|_{\xi,r} c_j|^2}\\
&= \sup_{|c|=1}|N_{\xi,r}(h)c| = |N_{\xi,r}(h)|\\
&= \|h\|_{\xi,r}\ ,
\end{aligned}
$$

where the inequality follows from the already proven scalar case (2.5.13). □

2.5.6. Majorants. Let us, now, recall briefly Cauchy's theory of majorants (see [**77**], chapter 5, for generalities).

Given two analytic functions f and F on $D^m_r(y_0)$ with (convergent) power series expansions

$$
f(y) = \sum_{k\in\mathbb{N}^m} f_k(y-y_0)^k\ , \qquad F(y) = \sum_{k\in\mathbb{N}^m} F_k(y-y_0)^k\ ,
$$

we say that F is a majorant for f, $f \prec F$, if $|f_k| \le F_k$ for any k.

Clearly, *majorization is preserved by sum and composition* (see [**77**]). Also, if $f \prec F$, then

$$
\|f\|_r \le F(y_0 + \underbrace{(r,\dots,r)}_{m\ \text{times}})\ . \qquad (2.5.14)
$$

For example the following elementary majorizations hold:

$$\sqrt{1-y} \prec 2-\sqrt{1-y}\ , \qquad (m=1\ ,\ y_0=0)\ ;$$
$$(1-y)^{-\frac{1}{2}} \prec (1-y)^{-\frac{1}{2}}\ , \qquad (m=1\ ,\ y_0=0)\ ;$$
$$y \prec |y_0|+y-y_0\ , \qquad (m=1\ , y_0\in\mathbb{C})\ ;$$
$$\frac{1}{y} \prec \frac{1}{|y_0|-(y-y_0)}\ , \qquad (m=1\ , y_0\in\mathbb{C}\backslash\{0\})\ ;$$
$$e(y_1,y_2) := \sqrt{1-\Big(\frac{y_2}{y_1}\Big)^2} \prec E(y_1,y_2) := 2-\sqrt{1-\Big(\frac{y_2}{2y_{01}-y_1}\Big)^2}\ ,$$
$$(m=2\ , y_0\in\mathbb{R}^2_+)\ ;$$

$$\frac{1}{e(y_1,y_2)} \prec \tilde{E}(y_1,y_2) := \Big(1-\Big(\frac{y_2}{2y_{01}-y_1}\Big)^2\Big)^{-\frac{1}{2}}\ , \qquad (m=2\ , y_0\in\mathbb{R}^2_+)\ . \tag{2.5.15}$$

Finally, we shall need the following simple result.

LEMMA 2.5.2. *Let $d, m, M\in\mathbb{Z}_+$ and let*

$$f(x,y) := \sum_{n\in\mathbb{Z}^d} f_n(y)\,\exp(in\cdot x)\ , \tag{2.5.16}$$

be a (absolutely) convergent series where $x\in\mathbb{T}^d$ and the f_n's are analytic functions on a complex ball around $y_0\in\mathbb{R}^m_+$. For $1\le j\le M$, let $a^{(j)}(\theta)$ and $b^{(j)}(\theta)$ be, respectively, $\mathbb{R}^d$-valued and $\mathbb{R}^m$-valued functions of $\theta\in\mathbb{T}^d$ and analytic on $\mathbb{T}^d_\xi$ for some $\xi>0$ and let

$$a(\theta,\varepsilon) := \sum_{j=1}^{M} a^{(j)}(\theta)\ \varepsilon^j\ , \qquad b(\theta,\varepsilon) := \sum_{j=1}^{M} b^{(j)}(\theta)\ \varepsilon^j\ .$$

Let F_n be a majorant of f_n:

$$f_n \prec F_n\ , \qquad \forall\ n\in\mathbb{Z}^d\ . \tag{2.5.17}$$

Let, also, for $1\le j\le M$ and $n\in\mathbb{Z}^d$, $A^{(j)}_n\in[0,\infty)$ and $B^{(j)}\in[0,\infty)^m$ be such that

$$\|n\cdot a^{(j)}\|_\xi \le A^{(j)}_n\ , \qquad \|b^{(j)}_i\|_\xi \le B^{(j)}_i\ , \qquad \forall\ 1\le j\le M\ , \quad \forall\ 1\le i\le m\ ,$$

and let

$$A_n(\varepsilon) := \sum_{j=1}^{M} A^{(j)}_n \varepsilon^j\ , \qquad B(\varepsilon) := \sum_{j=1}^{M} B^{(j)}\varepsilon^j\ .$$

Finally, define

$$\begin{aligned} \tilde{f}(\varepsilon;\theta) &:= f\big(\theta+a(\theta,\varepsilon), y_0+b(\theta,\varepsilon)\big)\ ,\\ \tilde{f}_M(\varepsilon;\theta) &:= \sum_{j\ge M+1} \big[\tilde{f}(\cdot,\theta)\big]_j \varepsilon^j\ ,\\ F(\varepsilon) &:= \sum_{n\in\mathbb{Z}^d} F_n\big(y_0+B(\varepsilon)\big)\ \exp\big(|n|\xi+A_n(\varepsilon)\big)\ , \end{aligned} \tag{2.5.18}$$

where,

$$[\,\cdot\,]_j := \frac{1}{j!}\frac{d^j}{d\varepsilon^j}\Big|_{\varepsilon=0}(\cdot)\ .$$

Then,

$$\|\big[\tilde{f}(\cdot,\theta)\big]_j\|_\xi \le \big[F\big]_j\ , \qquad \forall\ j\ge 0\ ; \tag{2.5.19}$$

$$\sup_{|\varepsilon|\le\varepsilon_0} \|\tilde{f}_M\|_\xi \le F(\varepsilon_0) - \sum_{j=0}^{M} \big[F\big]_j {\varepsilon_0}^j\ , \tag{2.5.20}$$

for any $\varepsilon_0>0$ for which $F(\varepsilon_0)<\infty$.

The same statement holds if f in (2.5.16) is replaced by

$$f(x,y) := \sum_{n\in\mathbb{Z}^d} f_n(y)\, c_n(x)\ , \tag{2.5.21}$$

where $c_n(x) = \cos(n\cdot x)$ or $c_n(x) = \sin(n\cdot x)$.

PROOF. Estimate (2.5.20) is an immediate consequence of (2.5.19). Also the final statement about (2.5.21) follows immediately by writing c_n as $\big(\exp(in\cdot x)+\exp(-in\cdot x)\big)/2$ or $\big(\exp(in\cdot x) - \exp(-in\cdot x)\big)/(2i)$ and applying twice the result for f as in (2.5.16).
It remains to prove (2.5.19). To simplify notations, we shall discuss in detail only the case $m = d = 1$, leaving the straightforward generalization to the higher dimensional case to the reader.
By definition, for any $j \geq 0$,

$$\Big\|\Big[\tilde f(\cdot,\theta)\Big]_j\Big\|_\xi \leq \sum_n \Big\|\Big[f_n(y_0+b)\exp(in\theta+ina)\Big]_j\Big\|_\xi ,$$

so that (2.5.19) will follows from

$$\begin{aligned}&\Big\|\Big[f_n(y_0+b)\exp(in\theta+ina)\Big]_j\Big\|_\xi\\ &\leq \Big[F_n(y_0+B)\exp\big(|n|\xi + A_n\big)\Big]_j\ ,\qquad \forall n\in\mathbb{Z}\ ,\ j\geq 0\ .\end{aligned} \tag{2.5.22}$$

For $j = 0$ (2.5.22) is immediately implied by (2.5.10) and (2.5.17). Let, now, $j \geq 1$ and let

$$f_n(y) := \sum_{k\geq 0} f_{n,k}\,(y-y_0)^k \prec \sum_{k\geq 0} F_{n,k}\,(y-y_0)^k =: F_n(y)\ .$$

Then[2.13]

$$\begin{aligned}&\Big\|\Big[f_n(y_0+b)\exp(in\theta+ina)\Big]_j\Big\|_\xi\\
&\leq \sum_{k,h}\Big\|\exp(in\theta)\frac{f_{n,k}}{h!}\Big[b^k(ina)^h\Big]_j\Big\|_\xi\\
&\leq \exp(|n|\xi)\sum_{k,h}\frac{F_{n,k}}{h!}\Big\|\Big[b^k(na)^h\Big]_j\Big\|_\xi\\
&\leq \exp(|n|\xi)\sum_{k,h}\frac{F_{n,k}}{h!}\Bigg\|\sum_{\mathcal{I}_{hkj}} b^{(j_1)}\cdots b^{(j_k)}\big(na^{(\ell_1)}\big)\cdots\big(na^{(\ell_h)}\big)\Bigg\|_\xi\\
&\leq \exp(|n|\xi)\sum_{k,h}\frac{F_{n,k}}{h!}\sum_{\mathcal{I}_{hkj}}\|b^{(j_1)}\|_\xi\cdots\|b^{(j_k)}\|_\xi\,\|na^{(\ell_1)}\|_\xi\cdots\|na^{(\ell_h)}\|_\xi\\
&\leq \exp(|n|\xi)\sum_{k,h}\frac{F_{n,k}}{h!}\sum_{\mathcal{I}_{hkj}} B^{(j_1)}\cdots B^{(j_k)}A_n^{(\ell_1)}\cdots A_n^{(\ell_h)}\\
&= \exp(|n|\xi)\sum_{k,h}\frac{F_{n,k}}{h!}\Big[B^kA_n^h\Big]_j\\
&= \exp(|n|\xi)\Big[\sum_{k,h}\frac{F_{n,k}}{h!}B^kA_n^h\Big]_j\\
&= \exp(|n|\xi)\ \big[F_n(y_0+B)\ \exp(A_n)\big]_j\ ,\end{aligned}$$

where $\mathcal{I}_{hkj}$ means $j_1 + \cdots + j_k + \ell_1 + \cdots + \ell_h = j$. □

[2.13] The dumb indices k, h run over $\mathbb{N}$, while the indices j_r and ℓ_r run over $\mathbb{Z}_+$.

2.5.7. Small divisor estimates. Let $\omega \in \mathbb{R}^d$ be Diophantine (see (2.1.3)). For $p \in \mathbb{N}$, $k \in \mathbb{N}^d$ and $\delta > 0$, we define[2.14]

$$s_{p,k}(\delta;\omega) = s_{p,k}(\delta) := \sup_{n\in\mathbb{Z}^d/\{0\}} \Big(|n^k| \exp(-\delta|n|)|\omega\cdot n|^{-p}\Big) ;$$

$$s_{p,1}(\delta;\omega) = s_{p,1}(\delta) := \sqrt{\sum_{j=1}^d s_{p,e_j}(\delta)^2} . \tag{2.5.23}$$

Notice that (2.1.3) implies that, for $k \in \mathbb{N}^d$,

$$s_{p,k}(\delta) \le \Big(\frac{b}{e}\Big)^b \gamma^{-p}\delta^{-b} , \qquad b := |k|_1 + p\tau ;$$

indeed:

$$\begin{aligned} s_{p,k}(\delta) &\le \sup_{n\neq 0} \gamma^{-p}|n^k|\, |n|^{p\tau} \exp(-\delta|n|) \\ &\le \sup_{n\neq 0} \gamma^{-p}|n|^{|k|_1} |n|^{p\tau} \exp(-\delta|n|) \\ &\le \gamma^{-p}\delta^{-b} \sup_{t>0} \Big(t^b \exp(-t)\Big) \\ &= \gamma^{-p}\delta^{-b}\Big(\frac{b}{e}\Big)^b . \end{aligned}$$

Let $f \in \mathcal{R}(\mathbb{T}^d_\xi, X_q)$, let $p \in \mathbb{N}$ and $0 < \delta \le \xi$. Then

$$\|D^{-p} f\|_{\xi-\delta} \le s_{p,0}(\delta)\ \|f\|_\xi , \tag{2.5.24}$$

$$\|D^{-p}\partial_\theta f\|_{\xi-\delta} \le s_{p,1}(\delta)\ \|f\|_\xi ; \tag{2.5.25}$$

in (2.5.24) f is assumed to have zero average over $\mathbb{T}^d$.

Below, we shall only use (2.5.24) with $q = 1$ (i.e., with f being $\mathbb{C}^m$-valued) and (2.5.25) with $p = 0$ and $q = 1, 2$ and with $p = 2$ and f scalar.

The proof of (2.5.24) and (2.5.25) follows easily from the definitions given; as an example let us check (2.5.25) in the case $f \in \mathcal{R}(\mathbb{T}^d_\xi, \mathbb{C})$ and $f \in \mathcal{R}(\mathbb{T}^d_\xi, \mathbb{C}^m)$. If $f \in \mathcal{R}(\mathbb{T}^d_\xi, \mathbb{C})$, we find

$$\begin{aligned} \|D^{-p}\partial_\theta f\|_{\xi-\delta} &= \|\sum_{n\neq 0} \frac{in}{(i\omega\cdot n)^p} f_n \exp(in\cdot x)\|_{\xi-\delta} \\ &= |N_{\xi-\delta}\Big(\sum \frac{in}{(i\omega\cdot n)^p} f_n \exp(in\cdot x)\Big)| \\ &= \sqrt{\sum_j \|\sum_n \frac{in_j}{(i\omega\cdot n)^p} f_n \exp(in\cdot x)\|^2_{\xi-\delta}} \\ &= \sqrt{\sum_j \Big(\sum_n \frac{|n_j|}{|\omega\cdot n|^p}|f_n|\ \exp(|n|(\xi-\delta))\Big)^2} \\ &\le \sqrt{\sum_j s_{p,e_j}(\delta)^2 \|f\|^2_\xi} \\ &= s_{p,1}(\delta)\ \|f\|_\xi ; \end{aligned} \tag{2.5.26}$$

[2.14] $\{e_j\}$ denotes the standard orthonormal basis $e_1 := (1, 0, 0, ..., 0)$, $e_2 := (0, 1, 0, ..., 0)$, ...

if $f \in \mathcal{R}(\mathbb{T}^d_\xi, \mathbb{C}^m)$,

$$\begin{aligned}
\|D^{-p}\partial_\theta f\|_{\xi-\delta} &= |N_{\xi-\delta}(D^{-p}\partial_\theta f)| \\
&= \sup_{|c|=1} |N_{\xi-\delta}(D^{-p}\partial_\theta f)\ c| \\
&= \sup_{|c|=1} \sqrt{\sum_i |\sum_j \|D^{-p}\partial_{\theta_j} f_i\|_{\xi-\delta}\ c_j|^2} \\
&\le \sup_{|c|=1} \sqrt{\sum_i (\sum_j s_{p,e_j}(\delta)\ c_j)^2 \|f_i\|_\xi^2} \\
&\le s_{p,1}(\delta)\ \|f\|_\xi\ ,
\end{aligned}$$

where in the first inequality we used (2.5.26). The estimates discussed in this paragraph in the context of supremum norms are more delicate (see [**121**]).

2.6. The KAM Norm Map

The purpose of this section is to *equip with* (careful) *estimates* the functional KAM map defined in section 2.4, proving, in particular, the existence, under suitable assumptions, of $(c, a) \in \mathbb{R}^d \times \mathbb{R}$ satisfying (2.3.14).

We begin by introducing a set of parameters controlling the norms of relevant quantities associated to the approximate torus (u, v, ω).

Assume that H is defined and bounded on $\mathbb{T}^d_{\bar\xi} \times D^d_r(y_0)$ for some $\bar\xi, r > 0$ and $y_0 \in \mathbb{R}^d$. Let $E_{p,q}$ be positive numbers such that[2.15]

$$\begin{aligned}
&\|H_x\|_{\bar\xi,r} \le E_{1,0}\ , && \|H_y\|_{\bar\xi,r} \le E_{0,1}\ , \\
&\|H_{xy}\|_{\bar\xi,r} \le E_{1,1}\ , && \|H_{yy}\|_{\bar\xi,r} \le E_{0,2}\ , \\
&\|H_{xxy}\|_{\bar\xi,r} \le E_{2,1}\ , && \|H_{xyy}\|_{\bar\xi,r} \le E_{1,2}\ , \\
&\|H_{xxx}\|_{\bar\xi,r} \le E_{3,0}\ , && \|H_{yyy}\|_{\bar\xi,r} \le E_{0,3}\ .
\end{aligned} \tag{2.6.1}$$

Fix $\rho > 0$ and recall the definitions given in Proposition 2.3.5 and the definition of $\mathcal{A}$ given in (2.3.35).

Let γ be as in (2.1.3), *let $\Omega \ge |\omega|$ and let ξ, F, G, $\bar h$, M, $\overline{M}$, U, V, $\tilde V$, $\overline{A}$ be non negative numbers such that*

$$\begin{aligned}
&\|f\|_\xi \le F\ , && \|g\|_\xi \le G\ , \quad |h| \le \bar h\ , \\
&\|\mathcal{M}\|_\xi \le M\ , && \|\mathcal{M}^{-1}\|_\xi \le \overline{M}\ , \\
&\sup_{\mathbb{T}^d_\xi} |\operatorname{Im}\ u| \le U\ , && \|v\|_\xi \le V\ , \\
&\|v_\theta\|_\xi \le \tilde V\ , && |\mathcal{A}^{-1}| \le \overline{A}\ ,
\end{aligned} \tag{2.6.2}$$

and such that

$$\sup_{\mathbb{T}^d_\xi} |\operatorname{Im}\ u| \le U < \bar\xi - \xi\ , \qquad \sup_{\mathbb{T}^d_\xi} |v(\theta) - y_0|_\infty < r. \tag{2.6.3}$$

Fix

$$1 < \kappa \le 2\ , \qquad 0 < \delta < \frac{\xi}{2}\ . \tag{2.6.4}$$

[2.15] Recall that by our definitions $\|H_{xy}\|_{\bar\xi,r} = \|H_{yx}\|_{\bar\xi,r}$ and $\|H_{xxy}\|_{\bar\xi,r} = \|H_{xyx}\|_{\bar\xi,r} = \|H_{yxx}\|_{\bar\xi,r}$ (and similarly for $\|H_{yxx}\|_{\bar\xi,r}$).

Finally, let $\sigma_{p,k}$ be upper bounds (computable in a finite number of steps) on[2.16] *$s_{p,k}$ defined in (2.5.23), § 2.5.7:*

$$s_{p,k}(\delta;\omega) \le \sigma_{p,k}(\delta) = \sigma_{p,k}(\delta;\omega)\ . \tag{2.6.5}$$

We shall *define the KAM norm map* as the map

$$\begin{aligned} \widehat{\mathcal{K}}_\delta := \widehat{\mathcal{K}}_{\delta,\sigma_{p,k}} : \quad & (\xi,\gamma,F,G,\bar{h},M,\overline{M},U,V,\tilde{V},\overline{A}) \mapsto \\ & (\xi',\gamma',F',G',\bar{h}',M',\overline{M}',U',V',\tilde{V}',\overline{A}')\ , \end{aligned} \tag{2.6.6}$$

where, of course, the primed quantities refer to the new approximate torus defined in Proposition 2.3.5, so that $|\omega'\cdot n| \ge \gamma'|n|^{-\tau}$, $\|f'\|_{\xi'} \le F'$, etc.

The rest of this section is devoted to the computation of $\widehat{\mathcal{K}}$ (which means to compute $\xi',\gamma',F',\ldots$ so that $(2.6.2)'$ holds for the primed quantities).

First of all, we define ξ' as

$$\xi' := \xi - 2\delta\ .$$

In the following bounds, we shall use systematically the general properties discussed in § 2.5 (see, in particular, § 2.5.4, 2.5.5 and 2.5.7); η_k will denote positive numbers "proportional" to the size of the error functions f and g or to the "energy error" h.

We start by estimating the quantities appearing in Lemma 2.3.9 in order to ensure (using Lemma 2.3.7) the existence of (c,a) satisfying (2.3.14).

Here is the first list of estimates that the reader will check without difficulty[2.17]:

$$\begin{aligned}
\|b\|_\xi &\le \tilde{V}F + MG =: \eta_1\ ,\\
\|D^{-1}b\|_{\xi-\delta'} &\le \sigma_{1,0}(\delta')\|b\|_\xi\\
&\le \sigma_{1,0}(\delta')(\tilde{V}F+MG)\ , \qquad \forall 0<\delta'\le\xi\ ,\\
|D^{-1}b(\theta)| &\le \|D^{-1}b\|_0\\
&\le \sigma_{1,0}(\xi)(\tilde{V}F+MG) =: \eta_2\ , \qquad \forall\theta\in\mathbb{T}^d\ ,\\
|\langle\mathcal{M}^{-1}f\rangle| &\le \overline{M}F =: \eta_3\ ,\\
|\mathcal{T}|_\xi &\le \overline{M}^2 E_{0,2}\ ,\\
|\langle\mathcal{T}D^{-1}b\rangle &\le \overline{M}^2 E_{0,2}\eta_2 =: \eta_4\ ,\\
|D^{-1}(\partial_\theta H^0)|_0 &= |D^{-1}\partial_\theta D^{-1}(H^0_x\cdot f + H^0_y\cdot g)|_0\\
&= |\partial_\theta D^{-2}(H^0_x\cdot f + H^0_y\cdot g)|_0\\
&\le \sigma_{2,1}(\xi)(E_{1,0}F+E_{0,1}G) =: \eta_5\ ,\\
|\langle\mathcal{T}D^{-1}(\partial_\theta H^0)\rangle| &\le \overline{M}^2 E_{0,2}\eta_5 =: \eta_6\ ,\\
|v'(0;c,a)-v'(0;0,0)| &\le \overline{M}(|c| + \frac{|a|}{1-|a|}\eta_5)\ , \qquad \forall |a|<1\ .
\end{aligned} \tag{2.6.7}$$

In order to guarantee that $v(0) + t\mathcal{M}^{-T}(0)D^{-1}b(0)$ lies, for $t\in[0,1]$, inside the y-domain of definition of H *we assume that*

$$|v(0)-y_0|_\infty + \overline{M}\eta_2 \le r\ . \tag{2.6.8}$$

[2.16] In fact, we shall need only $\sigma_{1,0}$ and $\sigma_{p,1}$ with $p=0,1,2$.

[2.17] Recall that $b := b_0 = v_\theta^T f - \mathcal{M}^T g$ and that $\langle b\rangle = 0$.

If such condition holds, we find (for a suitable $t_1 \in (0,1)$)

$$\begin{aligned} &|H_y(0,v_0') - H_y(0,v(0))| \\ &\quad = |H_{yy}\Big(0, v(0) + t_1 \mathcal{M}^{-T}(0) D^{-1} b(0)\Big) \mathcal{M}^{-T}(0) D^{-1} b(0)| \\ &\quad \le \overline{M} E_{0,2} \eta_2 \ . \end{aligned} \tag{2.6.9}$$

We can now estimate the norm of the matrix $\mathcal{N}$ defined in (2.3.29). Let

$$\mathcal{N} =: \frac{1}{\rho} \begin{pmatrix} 0 & \mathcal{N}_{12} \\ \mathcal{N}_{21} & \dfrac{\mathcal{N}_{22}}{\rho} \end{pmatrix} , \qquad \mathcal{N}_{12}, \mathcal{N}_{21} \in \mathbb{R}^d , \qquad \mathcal{N}_{22} \in \mathbb{R} .$$

By (2.6.7) and (2.6.9), we find

$$\begin{aligned} |\mathcal{N}_{12}| &\le \eta_3 + \eta_6 , \\ |\mathcal{N}_{21}| &\le \overline{M}^2 E_{0,2} \eta_2 , \\ |\mathcal{N}_{22}| &\le \overline{M} E_{0,1} \eta_5 , \end{aligned}$$

so that, by (2.5.4), we obtain

$$|\mathcal{N}| \le \frac{1}{\rho} \max\{N_1 \ , N_2\} =: \eta_7 , \tag{2.6.10}$$

where

$$\begin{aligned} N_1 &:= \sqrt{[\overline{M}^2 E_{0,2}\eta_2]^2 + \frac{\overline{M}^3 E_{0,2}\eta_2 E_{0,1}\eta_5}{\rho}} , \\ N_2 &:= \sqrt{(\eta_3+\eta_6)^2 + \Big[\frac{\overline{M} E_{0,1}\eta_5}{\rho}\Big]^2 + \frac{\overline{M}^3 E_{0,2}\eta_2 E_{0,1}\eta_5}{\rho}} . \end{aligned} \tag{2.6.11}$$

Recall that

$$F'(0) = \mathcal{A} + \mathcal{N}$$

(compare (2.3.29) and (2.3.35)). Then

$$|F'(0)^{-1}| = |(I + \mathcal{A}^{-1}\mathcal{N})^{-1}\mathcal{A}^{-1}| \le |\mathcal{A}^{-1}|(1 - |\mathcal{A}^{-1}||\mathcal{N}|)^{-1} \le \overline{A}(1 - \overline{A}|\mathcal{N}|)^{-1} ,$$

so that

$$|\mathcal{B}| := |F'(0)^{-1}| \le \frac{\overline{A}}{1 - \overline{A}\eta_7} =: B .$$

Next, define

$$\eta_8 := \sqrt{(\eta_3 + \eta_4)^2 + \frac{1}{\rho^2}(\overline{h} + E_{0,1}\overline{M}\eta_2)^2} ,$$

let κ be as in (2.6.4) and let

$$\eta_0 := \min\Big\{1, \frac{\kappa B \eta_8}{\rho}\Big\} . \tag{2.6.12}$$

Let, also,

$$\Omega(\rho\eta_0) := \{(c,a) : \ |c| \le \rho\eta_0 \ , \ |a| \le \eta_0\} ,$$

and *assume that*

$$|v(0) - y_0|_\infty + \overline{M}(\eta_2 + \rho\eta_0 + \frac{\eta_0\eta_5}{1-\eta_0}) \le r . \tag{2.6.13}$$

Notice that (2.6.13) *implies* (2.6.8) *and it implies that* $v_0' + t(v'(0) - v_0')$ *is inside the set* $\{|y - y_0|_\infty < r\}$, *for all* $0 \le t \le 1$. Furthermore, one sees easily that

$$
\begin{aligned}
|F(0)| &\le \eta_8 \,, \\
|v'(0; c, a) - v_0'| &\le \overline{M}(\rho\eta_0 + \frac{\eta_0\eta_5}{1-\eta_0}) =: \eta_9 \,, \\
\sup_{\Omega(\rho\eta_0)} |\widetilde{F}(\alpha)| &\le \sqrt{\Big(\frac{\eta_0^2\eta_6}{1-\eta_0}\Big)^2 + \Big(\frac{E_{0,2}\eta_9^2}{2\rho} + \frac{\eta_0^2}{1-\eta_0}\frac{E_{0,1}\overline{M}\eta_5}{\rho}\Big)^2} \\
&=: \eta_{10}^2 \,, \\
|\delta_H| &\le E_{0,2}\eta_9 \,, \\
\sup_{\Omega(\rho\eta_0)} |\widetilde{F}_\alpha(\alpha)| &\le \Bigg[\max\Bigg\{\Big(\frac{E_{0,2}\eta_9\overline{M}}{\rho}\Big)^2 \,,\, \Big(\frac{2+\eta_0}{(1-\eta_0)^2}\frac{\eta_0\eta_6}{\rho}\Big)^2 \\
&\quad + \Big(\frac{E_{0,2}\eta_9}{\rho^2} + \eta_0\frac{2+\eta_0}{(1-\eta_0)^2}\frac{E_{0,1}}{\rho^2}\Big)^2 \overline{M}^2\eta_5^2\Bigg\} \\
&\quad + \frac{E_{0,2}\eta_9\overline{M}}{\rho}\Big(\frac{E_{0,2}\eta_9}{\rho^2} + \eta_0\frac{2+\eta_0}{(1-\eta_0)^2}\frac{E_{0,1}}{\rho^2}\Big)\overline{M}\eta_5\Bigg]^{\frac{1}{2}} \\
&=: \eta_{11} \,, \\
\sup_{\Omega(\rho\eta_0)} |\alpha - \mathcal{B}F(\alpha)| &\le B(\eta_8 + \eta_{10}^2) =: \eta_{12} \,, \\
\sup_{\Omega(\rho\eta_0)} |I - \mathcal{B}F'(\alpha)| &\le B\eta_{11} \,.
\end{aligned}
$$

The conditions for applying Lemma 2.3.7 are implied by

$$\eta_{12} \le \rho\eta_0 \,, \qquad B\eta_{11} < 1 \,,$$

and one sees that $\eta_{12} \le \rho\eta_0$ is implied by $\eta_{10}^2 \le (\kappa - 1)\eta_8$ so that conditions for applying Lemma 2.3.7 are, now, implied by

$$\eta_{10}^2 \le (\kappa - 1)\eta_8 \,, \qquad B\eta_{11} < 1 \,. \tag{2.6.14}$$

As a corollary of Lemma 2.3.7, we thus obtain the following

LEMMA 2.6.1 (On the solution of (2.3.14)). *Assume* (2.6.14). *Then, one has*

$$\sup_{|\alpha| \le \rho\eta_0} |(F'(\alpha))^{-1}| \le B(1 - B\eta_{11})^{-1}$$

and there exists a unique solution $(c_0, \hat{a}_0)$ *of*

$$F(c_0, \hat{a}_0) = 0 \,, \tag{2.6.15}$$

i.e., of (2.3.14). *Furthermore,* $(c_0, \hat{a}_0)$ *satisfies*

$$\max\Big\{|c_0| \,,\, |\hat{a}_0|\Big\} \le |(c_0, \hat{a}_0)| \le \frac{B}{1 - B\eta_{11}}\eta_8 =: \eta_{13} \,. \tag{2.6.16}$$

From here on, $\hat{a}$, a and c stands, respectively, for $\hat{a}_0$,

$$a_0 := \frac{\hat{a}_0}{\rho} \tag{2.6.17}$$

and c_0 as in (2.6.15), (2.6.16).

We now proceed to estimate the new Diophantine constant γ'; let

$$\omega' := \omega_a = (1+a)\omega$$

with $a = a_0$ as in (2.6.15) and (2.6.16). Then, ω' verifies

$$|\omega' \cdot n| \geq \frac{\gamma'}{|n|^\tau} \qquad \forall n \in \mathbb{Z}^d/\{0\}$$

with

$$\gamma' := \Big(1 - \frac{\eta_{13}}{\rho}\Big)\gamma\ ,$$

provided

$$\eta_{13} < \rho\ . \tag{2.6.18}$$

Next, recall the definition of z:

$$z := \mathcal{M}D_{\omega_a}^{-1}\hat{z} + \mathcal{M}\hat{z}_0$$

where

$$\begin{aligned}
\hat{z} &:= \mathcal{T}c + \mathcal{T}D_{\omega_a}^{-1}b_a - \mathcal{M}^{-1}f_a\ ,\\
\hat{z}_0 &:= -(D_{\omega_a}^{-1}\hat{z})(0)\ ,\\
f_a &:= (1+a)f + aH_y^0\ ,\\
g_a &:= (1+a)g - aH_x^0\ ,\\
b_a &:= v_\theta^T f_a - \mathcal{M}^T g_a\ ,\\
D_{\omega_a}^{-1}b_a &= D^{-1}b + aD_{\omega_a}^{-1}(\partial_\theta H^0)\ ,\\
\partial_\theta H^0 &= \partial_\theta D^{-1}(H_x^0 \cdot f + H_y^0 \cdot g)\ ,
\end{aligned}$$

(recall that $\langle H_x^0 \cdot f + H_y^0 \cdot g\rangle = 0$).

Then, the following bounds are easily checked:

$$
\begin{aligned}
|f_a|_\xi &\le \Big(1+\frac{\eta_{13}}{\rho}\Big)F+\frac{\eta_{13}}{\rho}E_{0,1} =: \eta_{14} \ ,\\
|g_a|_\xi &\le \Big(1+\frac{\eta_{13}}{\rho}\Big)G+\frac{\eta_{13}}{\rho}E_{1,0} =: \eta_{15} \ ,\\
|b_a|_\xi &\le \tilde V\eta_{14}+M\eta_{15} =: \eta_{16} \ ,\\
|\mathcal{T}c|_\xi &\le \overline{M}^2 E_{0,2}\eta_{13} =: \eta_{17} \ ,\\
|D^{-1}b|_{\xi-\delta} &:= |D^{-1}(v_\theta f-\mathcal{M}^T g)|_{\xi-\delta}\le \sigma_{1,0}(\delta)(|v_\theta f|_\xi+|\mathcal{M}^T g|_\xi)\\
&\le \sigma_{1,0}(\delta)(\tilde V F+MG)=\sigma_{1,0}(\delta)\eta_1 =: \eta_{18} \ ,\\
|D^{-1}(\partial_\theta H^0)|_{\xi-\delta} &= |\partial_\theta D^{-2}(H^0_x\cdot f+H^0_y\cdot g)|_{\xi-\delta}\le \sigma_{2,1}(\delta)(E_{1,0}F+E_{0,1}G)\\
&=: \eta_{19} \ ,\\
|D^{-1}_{\omega_a}b_a|_{\xi-\delta} &\le \eta_{18}+\frac{1}{\rho}\,\frac{\eta_{13}\eta_{19}}{1-(\eta_{13}/\rho)} =: \eta_{20} \ ,\\
|\hat z|_{\xi-\delta} &\le \eta_{17}+\overline{M}^2E_{0,2}\eta_{20}+\overline{M}\eta_{14} =: \eta_{21} \ ,\\
|\hat z_0| &= |(D^{-1}_{\omega_a}\hat z)(0)|\le |\frac{1}{1+a}||D^{-1}\hat z|_0\le \frac{1}{1-(\eta_{13}/\rho)}\sigma_{1,0}(\xi-\delta)\eta_{21}\\
&=: \eta_{22} \ ,\\
|z|_{\xi-2\delta} &\le M(|D^{-1}_{\omega_a}\hat z|_{\xi-2\delta}+\eta_{22})\le M(\frac{1}{1-(\eta_{13}/\rho)}|D^{-1}\hat z|_{\xi-2\delta}+\eta_{22})\\
&\le M(\frac{1}{1-(\eta_{13}/\rho)}\sigma_{1,0}(\delta)\eta_{21}+\eta_{22}) =: \eta_{23} \ , \qquad (2.6.19)\\
|z_\theta|_{\xi-2\delta} &\le |\mathcal{M}_\theta D^{-1}_{\omega_a}\hat z|_{\xi-2\delta}+|\mathcal{M}D^{-1}_{\omega_a}\partial_\theta\hat z|_{\xi-2\delta}+|\mathcal{M}_\theta\hat z_0|_{\xi-2\delta}\\
&\le M\sigma_{0,1}(2\delta)|D^{-1}_{\omega_a}\hat z|_{\xi-2\delta}+M|D^{-1}_{\omega_a}\partial_\theta\hat z|_{\xi-2\delta}+M\sigma_{0,1}(2\delta)\eta_{22}\\
&\le M\sigma_{0,1}(2\delta)\frac{1}{1-(\eta_{13}/\rho)}\sigma_{1,0}(\delta)\eta_{21}+M\sigma_{1,1}(\delta)\eta_{21}\\
&+ M\sigma_{0,1}(2\delta)\eta_{22}\\
&=: \eta_{24} \ .
\end{aligned}
$$

Recalling the definition of w:

$$w := \mathcal{M}^{-T}(v_\theta^T z+c+D^{-1}_{\omega_a}b_a) \ ,$$

we find

$$|w|_{\xi-2\delta}\le \overline{M}\ (\tilde V\eta_{23}+\eta_{13}+\eta_{20}) =: \eta_{25} \qquad (2.6.20)$$

and

$$
\begin{aligned}
|w_\theta|_{\xi-2\delta} &= \Big|(\mathcal{M}^{-T})_\theta\ (v_\theta^T z+c+D^{-1}_{\omega_a}b_a)+\mathcal{M}^{-T}\Big((v_\theta^T)_\theta z+v_\theta^T z_\theta+D^{-1}_{\omega_a}\partial_\theta b_a\Big)\Big|\\
&\le \sigma_{0,1}(2\delta)\eta_{25}+\overline{M}\Big(\tilde V\sigma_{0,1}(2\delta)\eta_{23}+\tilde V\eta_{24}+\eta_{20}\sigma_{0,1}(\delta)\Big) =: \eta_{26} \ .
\end{aligned}
$$

REMARK 2.6.2. The estimate on $|D^{-1}_{\omega_a}\partial_\theta b_a|_{\xi-2\delta}$ could be slightly improved, since

$$D^{-1}_{\omega_a}\partial_\theta b_a = D^{-1}\partial_\theta b + a\ D^{-1}_{\omega_a}\partial^2_\theta H^0 \ .$$

We proceed to the estimate the "quadratic" functions Q_i. Concerning Q_1 and Q_2 we have:

$$\begin{aligned} |Q_1|_{\xi-2\delta} &:= |H_y(\theta+u+z, v+w) - H_y^0 - H_{yx}^0 z - H_{yy}^0 w|_{\xi-2\delta} \\ &\leq \frac{1}{2}E_{2,1}\eta_{23}^2 + E_{1,2}\eta_{23}\eta_{25} + \frac{1}{2}E_{0,3}\eta_{25}^2 =: q_1 \ , \\ |Q_2|_{\xi-2\delta} &\leq \eta_{14}\sigma_{0,1}(2\delta)\overline{M}\eta_{23} =: q_2 \ . \end{aligned}$$

Moreover, one has

$$\begin{aligned} \left|D_{\omega_a}^{-1}\mathcal{G}_a\right|_{\xi-2\delta} &\leq \frac{2}{1-(\eta_{13}/\rho)}\ \sigma_{1,0}(\delta)\left|v_\theta^T\partial_\theta f_a - \mathcal{M}^T\partial_\theta g_a\right|_{\xi-\delta} \\ &\leq \frac{2}{1-(\eta_{13}/\rho)}\ \sigma_{1,0}(\delta)\ (\tilde{V}\sigma_{0,1}(\delta)\eta_{14} + M\sigma_{0,1}(\delta)\eta_{15}) \\ &=: \eta_{27} \ , \end{aligned}$$

and

$$|Q_3|_{\xi-2\delta} \leq E_{0,2}\overline{M}^2\eta_{23}\eta_{27} =: q_3 \ .$$

Thus,

$$|f'|_{\xi-2\delta} \leq q_1 + q_2 + q_3 =: F' \ .$$

Concerning g' one obtains:

$$\begin{aligned} |Q_4|_{\xi-2\delta} &\leq \frac{1}{2}E_{3,0}\eta_{23}^2 + E_{2,1}\eta_{23}\eta_{25} + \frac{1}{2}E_{1,2}\eta_{25}^2 =: q_4 \ , \\ |Q_5|_{\xi-2\delta} &\leq \overline{M}\sigma_{0,1}(2\delta)\eta_{15}\eta_{23} =: q_5 \ , \\ |Q_6|_{\xi-2\delta} &\leq \overline{M}\sigma_{0,1}(2\delta)\eta_{14}\eta_{25} =: q_6 \ . \end{aligned}$$

Thus,

$$|g'|_{\xi-2\delta} \leq q_4 + q_5 + q_6 + \overline{M}\tilde{V}(q_2+q_3) =: G' \ .$$

Notice that, since we solved exactly, by the Lemma 2.3.7, the equation $F(c,\hat{a}) = 0$ we have

$$h' \ = \ 0 \ .$$

Next,

$$|\mathcal{M}'|_{\xi'} := |\mathcal{M} + z_\theta|_{\xi'} \leq M + \eta_{24} \ ,$$

namely

$$M' \ := M + \eta_{24} \ .$$

From

$$\sup_{\mathbb{T}^d_{\xi'}} |\operatorname{Im} u'| = \sup_{\mathbb{T}^d_{\xi'}} |\operatorname{Im}(u+z)| \leq U + \eta_{23} \ ,$$

it follows that we can take

$$U' := U + \eta_{23} \ .$$

From

$$\|v'\|_{\xi'} = \|v+w\|_{\xi'} \leq V + \eta_{25} \ ,$$

it follows that we can take

$$V' := V + \eta_{25} \ .$$

From

$$\|v'_\theta\|_{\xi'} = \|v_\theta + w_\theta\|_{\xi'} \leq \tilde{V} + \eta_{26} \ ,$$

it follows that

$$\tilde{V}' := \tilde{V} + \eta_{26} \ .$$

REMARK 2.6.3. In general, if A is invertible and $|B|$ is small, i.e. $|B|\ |A|^{-1} < 1$, then

$$(A+B)^{-1} = A^{-1} + C$$

with

$$C := [(I + A^{-1}B)^{-1} - I]A^{-1} = \sum_{j\geq 1} (A^{-1}B)^j\ A^{-1}$$

and

$$|C| \leq \frac{|A^{-1}|^2|B|}{1 - |A^{-1}|\ |B|} \ .$$

Thus, concerning $(\mathcal{M}')^{-1} := (\mathcal{M} + z_\theta)^{-1}$ we find

$$(\mathcal{M}')^{-1} = \mathcal{M}^{-1} + \Big((I + \mathcal{M}^{-1}z_\theta)^{-1} - I\Big)\mathcal{M}^{-1} =: \mathcal{M}^{-1} + \mathcal{C}_1$$

and

$$\|\mathcal{C}_1\|_{\xi'} \leq \frac{\overline{M}^2 \eta_{24}}{1 - \overline{M}\eta_{24}} =: \eta_{28} \ ,$$

provided

(2.6.21) $$\overline{M}\eta_{24} < 1 \ .$$

We can then set

$$\overline{M}' := \overline{M} + \eta_{28} \ .$$

Defining $\mathcal{C}_2$ by

$$H_{yy}(\theta + u + z, v + w) = H^0_{yy} + \mathcal{C}_2 \ ,$$

we have, by Lagrange's formula,

$$\|\mathcal{C}_2\|_{\xi'} \leq E_{1,2}\eta_{23} + E_{0,3}\eta_{25} =: \eta_{29} \ .$$

Defining $\mathcal{C}_3$ by setting

$$\mathcal{T}' = \mathcal{T} + \mathcal{C}_3 \ , \qquad \mathcal{C}_3 := (\mathcal{M}^{-1} + \mathcal{C}_1)(H^0_{yy} + \mathcal{C}_2)(\mathcal{M}^{-T} + \mathcal{C}_1^T) - \mathcal{T} \ ,$$

we find

$$\|\mathcal{C}_3\|_{\xi'} \ \leq \ \overline{M}^2\eta_{29} + 2E_{0,2}\overline{M}\eta_{28} + 2\overline{M}\eta_{28}\eta_{29} + E_{0,2}\eta_{28}^2 + \eta_{28}^2\eta_{29} =: \eta_{30} \ .$$

Define

$$\chi' := \frac{1}{\rho}\ \mathcal{M}'^{-1}H_y(\theta + u + z, v + w) =: \chi + \chi_1 \ ,$$

so that

$$\chi_1 = \frac{1}{\rho}\ \Big(\mathcal{M}^{-1}\bar{\delta} + \mathcal{C}_1(H_y^0)^T + \mathcal{C}_1\bar{\delta}\Big) \ ,$$

with

$$\bar{\delta} := H_y(\theta + u + z, v + w) - H_y^0 \ .$$

Then, one finds

$$\begin{aligned}\|\chi_1\|_{\xi'} \ &\leq \ \frac{1}{\rho}\Big[\overline{M}(E_{1,1}\eta_{23} + E_{0,2}\eta_{25}) + E_{0,1}\eta_{28} + \eta_{28}(E_{1,1}\eta_{23} + E_{0,2}\eta_{25})\Big] \\ &=: \ \eta_{31} \ .\end{aligned}$$

Concerning $\mathcal{A}'^{-1}$ we have

$$\begin{aligned} \mathcal{A}'^{-1} &:= \begin{pmatrix} \langle \mathcal{T}' \rangle & -\langle \chi' \rangle \\ \chi'(0)^T & 0 \end{pmatrix}^{-1} = \begin{pmatrix} \langle \mathcal{T} \rangle + \mathcal{C}_3 & -\langle \chi \rangle - \langle \chi_1 \rangle \\ \chi(0)^T + \chi_1(0)^T & 0 \end{pmatrix}^{-1} \\ &=: \mathcal{A}^{-1} + \overline{\mathcal{C}} \, . \end{aligned}$$

If

$$\overline{\mathcal{C}}_* := \begin{pmatrix} \mathcal{C}_3 & -\langle \chi_1 \rangle \\ \chi_1(0)^T & 0 \end{pmatrix} ,$$

then

$$\overline{\mathcal{C}} := \{ (I + \mathcal{A}^{-1} \, \overline{\mathcal{C}}_*)^{-1} - I \} \mathcal{A}^{-1} \, .$$

Therefore,

$$|\overline{\mathcal{C}}| \le \frac{\overline{A}^2 |\overline{\mathcal{C}}_*|}{1 - \overline{A} |\overline{\mathcal{C}}_*|} \le \frac{\overline{A}^2 \sqrt{\eta_{30}^2 + \eta_{31}^2 + \eta_{30}\eta_{31}}}{1 - \overline{A} \sqrt{\eta_{30}^2 + \eta_{31}^2 + \eta_{30}\eta_{31}}} =: \eta_{32} \, ,$$

provided

$$\overline{A} \sqrt{\eta_{30}^2 + \eta_{31}^2 + \eta_{30}\eta_{31}} < 1 \, . \tag{2.6.22}$$

We shall then set

$$\overline{A}' := \overline{A} + \eta_{32} \, .$$

The computation of the KAM norm map $\widehat{\mathcal{K}}_{\delta, \sigma_{p,k}}$ is completed and we have proven the following

PROPOSITION 2.6.4 (The KAM norm map). *Fix $E \in \mathbb{R}$ and let (u, v, ω) be an approximate KAM torus with u and v real-analytic on $\mathbb{T}^d_\xi$ and such that the matrix $\mathcal{A}$ in (2.3.35) is invertible on $\mathbb{T}^d_\xi$. Let $\bar{\xi} > \xi$, r and $E_{p,q}$ be as in (2.6.1); let γ be as in (2.1.3); let F, G, $\bar{h}$, M, $\overline{M}$, U, V, $\tilde{V}$, $\overline{A}$ be as in (2.6.2) and assume (2.6.3). Fix κ and δ as in (2.6.4) and let $\sigma_{p,k} = \sigma_{p,k}(\delta; \omega)$ be as in (2.6.5), (2.5.23). Define* *the following non-negative numbers*[2.18]*:*

$$\eta_2 := \sigma_{1,0}(\xi)(\tilde{V}F + MG)\,;$$
$$\eta_3 := \overline{M}F\,;$$
$$\eta_4 := \overline{M}^2 E_{0,2}\eta_2\,;$$
$$\eta_5 := \sigma_{2,1}(\xi)(E_{1,0}F + E_{0,1}G)\,;$$
$$\eta_6 := \overline{M}^2 E_{0,2}\eta_5\,;$$
$$\eta_7 := \frac{1}{\rho}\max\left\{\eta_7^{(a)}\,,\eta_7^{(b)}\right\},$$
$$\text{where } \eta_7^{(a)} := \sqrt{[\overline{M}^2 E_{0,2}\eta_2]^2 + \frac{\overline{M}^3 E_{0,2}\eta_2 E_{0,1}\eta_5}{\rho}}\,,$$
$$\eta_7^{(b)} := \sqrt{(\eta_3+\eta_6)^2 + \left[\frac{\overline{M}E_{0,1}\eta_5}{\rho}\right]^2 + \frac{\overline{M}^3 E_{0,2}\eta_2 E_{0,1}\eta_5}{\rho}}\,;$$
$$\eta_8 := \sqrt{(\eta_3+\eta_4)^2 + \frac{1}{\rho^2}(\overline{h} + E_{0,1}\overline{M}\eta_2)^2}\,;$$
$$B := \frac{\overline{A}}{1-\overline{A}\eta_7}\,;$$
$$\eta_0 := \min\left\{1, \frac{\kappa B\eta_8}{\rho}\right\};$$
$$\eta_9 := \overline{M}(\rho\eta_0 + \frac{\eta_0\eta_5}{1-\eta_0})\,;$$
$$\eta_{10} := \left(\left(\frac{\eta_0^2\eta_6}{1-\eta_0}\right)^2 + \left(\frac{E_{0,2}\eta_9^2}{2\rho} + \frac{\eta_0^2}{1-\eta_0}\,\frac{E_{0,1}\overline{M}\eta_5}{\rho}\right)^2\right)^{\frac{1}{4}}\,;$$
$$\eta_{11} := \left(\max\left\{\left(\frac{E_{0,2}\eta_9\overline{M}}{\rho}\right)^2,\right.\right.$$
$$\left.\left(\frac{2+\eta_0}{(1-\eta_0)^2}\,\frac{\eta_0\eta_6}{\rho}\right)^2 + \left(\frac{E_{0,2}\eta_9}{\rho^2} + \eta_0\frac{2+\eta_0}{(1-\eta_0)^2}\,\frac{E_{0,1}}{\rho^2}\right)^2\overline{M}^2\eta_5^2\right\}$$
$$\left.+\left(\frac{E_{0,2}\eta_9\overline{M}}{\rho}\right)\left(\frac{E_{0,2}\eta_9}{\rho^2} + \eta_0\frac{2+\eta_0}{(1-\eta_0)^2}\,\frac{E_{0,1}}{\rho^2}\right)\overline{M}\eta_5\right)^{\frac{1}{2}}\,;$$
$$\eta_{13} := \frac{B}{1-B\eta_{11}}\eta_8\,;$$
$$\eta_{14} := \left(1+\frac{\eta_{13}}{\rho}\right)F + \frac{\eta_{13}}{\rho}E_{0,1}\,;$$
$$\eta_{15} := \left(1+\frac{\eta_{13}}{\rho}\right)G + \frac{\eta_{13}}{\rho}E_{1,0}\,;$$
$$\eta_{17} := \overline{M}^2 E_{0,2}\eta_{13}\,;$$
$$\eta_{18} := \sigma_{1,0}(\delta)(\tilde{V}F + MG)\,;$$
$$\eta_{19} := \sigma_{2,1}(\delta)(E_{1,0}F + E_{0,1}G)\,;$$
$$\eta_{20} := \eta_{18} + \frac{1}{\rho}\,\frac{\eta_{13}\eta_{19}}{1-(\eta_{13}/\rho)}\,;$$
$$\eta_{21} := \eta_{17} + \overline{M}^2 E_{0,2}\eta_{20} + \overline{M}\eta_{14}\,;$$
$$\eta_{22} := \frac{1}{1-(\eta_{13}/\rho)}\sigma_{1,0}(\xi-\delta)\eta_{21}\,;$$
$$\eta_{23} := M(\frac{1}{1-(\eta_{13}/\rho)}\sigma_{1,0}(\delta)\eta_{21} + \eta_{22})\,;$$
$$\eta_{24} := M\sigma_{0,1}(2\delta)\frac{1}{1-(\eta_{13}/\rho)}\sigma_{1,0}(\delta)\eta_{21} + M\sigma_{1,1}(\delta)\eta_{21} + M\sigma_{0,1}(2\delta)\eta_{22}\,;$$

[2.18] The quantities η_1, η_{12} and η_{16} do not appear in the following list.

$$\eta_{25} := \overline{M}\ (\tilde{V}\eta_{23} + \eta_{13} + \eta_{20})\,;$$
$$\eta_{26} := \sigma_{0,1}(2\delta)\eta_{25} + \overline{M}\Big(\tilde{V}\sigma_{0,1}(2\delta)\eta_{23} + \tilde{V}\eta_{24} + \eta_{20}\sigma_{0,1}(\delta)\Big)\,;$$
$$\eta_{27} := \frac{2}{1-(\eta_{13}/\rho)}\ \sigma_{1,0}(\delta)\ (\tilde{V}\sigma_{0,1}(\delta)\eta_{14} + M\sigma_{0,1}(\delta)\eta_{15})\,;$$
$$\eta_{28} := \frac{\overline{M}^2\eta_{24}}{1-\overline{M}\eta_{24}}\,;$$
$$\eta_{29} := E_{1,2}\eta_{23} + E_{0,3}\eta_{25}\,;$$
$$\eta_{30} := \overline{M}^2\eta_{29} + 2E_{0,2}\overline{M}\eta_{28} + 2\overline{M}\eta_{28}\eta_{29} + E_{0,2}\eta_{28}^2 + \eta_{28}^2\eta_{29}\,;$$
$$\eta_{31} := \frac{1}{\rho}\Big[\overline{M}(E_{1,1}\eta_{23} + E_{0,2}\eta_{25}) + E_{0,1}\eta_{28} + \eta_{28}(E_{1,1}\eta_{23} + E_{0,2}\eta_{25})\Big]\,;$$
$$\eta_{32} := \frac{\overline{A}^2\sqrt{\eta_{30}^2 + \eta_{31}^2 + \eta_{30}\eta_{31}}}{1-\overline{A}\sqrt{\eta_{30}^2 + \eta_{31}^2 + \eta_{30}\eta_{31}}}\,;$$
$$q_1 := \frac{1}{2}E_{2,1}\eta_{23}^2 + E_{1,2}\eta_{23}\eta_{25} + \frac{1}{2}E_{0,3}\eta_{25}^2\,;$$
$$q_2 := \eta_{14}\sigma_{0,1}(2\delta)\overline{M}\eta_{23}\,;$$
$$q_3 := E_{0,2}\overline{M}^2\eta_{23}\eta_{27}\,;$$
$$q_4 := \frac{1}{2}E_{3,0}\eta_{23}^2 + E_{2,1}\eta_{23}\eta_{25} + \frac{1}{2}E_{1,2}\eta_{25}^2\,;$$
$$q_5 := \overline{M}\sigma_{0,1}(2\delta)\eta_{15}\eta_{23}\,;$$
$$q_6 := \overline{M}\sigma_{0,1}(2\delta)\eta_{14}\eta_{25}\,.$$

Assume, now, that

$$\overline{M}\Big(\eta_2 + \rho\eta_0 + \frac{\eta_0\eta_5}{1-\eta_0}\Big) \le r - |v(0) - y_0|_\infty\,, \tag{2.6.23}$$
$$\eta_{10}^2 \le (\kappa - 1)\eta_8\,, \tag{2.6.24}$$
$$B\eta_{11} < 1\,, \tag{2.6.25}$$
$$\eta_{13} < \rho\,, \tag{2.6.26}$$
$$\overline{M}\eta_{24} < 1\,, \tag{2.6.27}$$
$$\overline{A}\sqrt{\eta_{30}^2 + \eta_{31}^2 + \eta_{30}\eta_{31}} < 1\,, \tag{2.6.28}$$
$$\eta_{23} \le 2\delta\,, \tag{2.6.29}$$
$$\eta_{25} < r - \sup_{\mathbb{T}^d_\xi}|v(\theta) - y_0|_\infty\,. \tag{2.6.30}$$

and define the KAM norm map $\widehat{\mathcal{K}}_{\delta,\sigma_{p,k}}$ *in* (2.6.6) *by letting*

$$\widehat{\mathcal{K}}_{\delta,\sigma_{p,k}}: \quad \begin{cases} \xi' = \xi - 2\delta\,, \\ \gamma' = \left(1 - \frac{\eta_{13}}{\rho}\right)\gamma\,, \\ F' = q_1 + q_2 + q_3\,, \\ G' = q_4 + q_5 + q_6 + \overline{M}\tilde{V}(q_2 + q_3)\,, \\ \bar{h}' = 0\,, \\ M' = M + \eta_{24}\,, \\ \overline{M}' = \overline{M} + \eta_{28}\,, \\ U' = U + \eta_{23}\,, \\ V' = V + \eta_{25}\,, \\ \tilde{V}' = \tilde{V} + \eta_{26}\,, \\ \overline{A}' = \overline{A} + \eta_{32}\,. \end{cases} \tag{2.6.31}$$

Let $(u',v',\omega')=\mathcal{K}(u,v,\omega)$ *be as in* (2.4.4), *and let* f', g', h', $\mathcal{M}'$ *and* $\mathcal{A}'$ *be as in* (2.4.3), (2.4.1) *and* (2.4.3) *after having replaced* (u,v,ω) *with*[2.19] (u',v',ω'). *Then* ω' *is* (γ',τ)*-Diophantine and*

$$\begin{aligned}
&\|f'\|_{\xi'}\le F' , && \|g'\|_{\xi'}\le G' , && h'=0=\bar h' ,\\
&\|\mathcal{M}'\|_{\xi'}\le M' , && \|\mathcal{M}'^{-1}\|_{\xi'}\le \overline{M}' , &&\\
&\sup_{\mathbb{T}^d_{\xi'}}|\,\mathrm{Im}\ u'|\le U' , && \|v'\|_{\xi'}\le V' , &&\\
&\|v'_\theta\|_{\xi'}\le \tilde V' , && \|\mathcal{A}'^{-1}\|_{\xi'}\le \overline{A}' . &&
\end{aligned}\tag{2.6.32}$$

Furthermore

$$\sup_{\mathbb{T}^d_{\xi'}}|\,\mathrm{Im}\,u'|<\bar\xi-\xi' ,\tag{2.6.33}$$

$$\sup_{\mathbb{T}^d_{\xi'}}|v'(\theta)-y_0|_\infty<r .\tag{2.6.34}$$

Note that the conditions (2.6.23)÷(2.6.28) correspond to, respectively, (2.6.13) (which implies also (2.6.8)), (2.6.14), (2.6.18), (2.6.21), (2.6.22), which are all the conditions used in this section to derive the above estimates.

Condition (2.6.29) together with (2.6.31), (2.6.32) and (2.6.3) imply that

$$\sup_{\mathbb{T}^d_{\xi'}}|\,\mathrm{Im}\,u'|\le U+\eta_{23}<\bar\xi-\xi+\eta_{23}\le\bar\xi-\xi+2\delta=\bar\xi-\xi' ,$$

which proves (2.6.33).

Condition (2.6.30) together with (2.6.20) and (2.6.3) imply that

$$\sup_{\mathbb{T}^d_{\xi'}}\big|v'(\theta)-y_0\big|_\infty\le\sup_{\mathbb{T}^d_{\xi'}}\big|v(\theta)-y_0\big|_\infty+\|w\|_{\xi'}\le\sup_{\mathbb{T}^d_{\xi'}}\big|v(\theta)-y_0\big|_\infty+\eta_{25}<r ,$$

which proves (2.6.34). □

REMARK 2.6.5. (i) Relations (2.6.33) and (2.6.34) imply that, for $\theta\in\mathbb{T}^d_{\xi'}$, one has

$$\big(\theta+u'(\theta),v'(\theta)\big)\in\mathbb{T}^d_{\bar\xi}\times D^d_r(y_0) ,$$

which is inside the analyticity domain of H. Thus if (2.6.33) and (2.6.34) are satisfied one can apply again the KAM map $\mathcal{K}$ to (u',v',ω').

(ii) When one iterates the KAM maps, it is useful to simplify conditions (2.6.23) and (2.6.30), as we proceed to explain. Let, for $1\le i\le j$,

$$\big(u^{(i)},v^{(i)},\omega^{(i)}\big):=\mathcal{K}\big(u^{(i-1)},v^{(i-1)},\omega^{(i-1)}\big) ,$$

and let $f^{(i)}$, $g^{(i)}$, $h^{(i)}$, $\mathcal{M}^{(i)}$ and $\mathcal{A}^{(i)}$ be as in (2.4.3), (2.4.1) and (2.4.3) after having replaced (u,v,ω) with $(u^{(i-1)},v^{(i-1)},\omega^{(i-1)})$; let $F^{(i)}$,..., $\overline{A}^{(i)}$ be the corresponding bounds on the norms $\|f^{(i)}\|_{\xi_i}$,..., $|\overline{A}^{(i)}|$ with $\xi_i=\xi_{i-1}-2\delta_{i-1}$ (δ_i being assigned numbers such that $\xi_{i-1}-2\delta_{i-1}>0$ for $i\ge1$). Then, conditions

[2.19] In particular, $H^0_{\cdot}$ has to be replaced by $H_{\cdot}(\theta+u',v')$.

(2.6.23)÷(2.6.30), indexed by j, needed to construct and control the approximate torus $\big(u^{(j+1)}, v^{(j+1)}, \omega^{(j+1)}\big)$, are immediately seen to be implied by[2.20]

$$\begin{aligned}
&\overline{M}^{(j)}\Big(\eta_2^{(j)} + \rho\eta_0^{(j)} + \frac{\eta_0^{(j)}\eta_5^{(j)}}{1-\eta_0^{(j)}}\Big) + \sum_{i=0}^{j-1}\eta_{25}^{(i)} \le r - |v^{(0)}(0) - y_0|_\infty \ , \\
&\big(\eta_{10}^{(j)}\big)^2 \le (\kappa - 1)\eta_8^{(j)} \ , \\
&B\eta_{11}^{(j)} < 1 \ , \\
&\eta_{13}^{(j)} < \rho \ , \\
&\overline{M}^{(j)}\eta_{24}^{(j)} < 1 \ , \\
&\overline{A}^{(j)}\sqrt{\big(\eta_{30}^{(j)}\big)^2 + \big(\eta_{31}^{(j)}\big)^2 + \eta_{30}^{(j)}\eta_{31}^{(j)}} < 1 \ , \\
&\eta_{23}^{(j)} \le 2\delta_j \ , \\
&\sum_{i=0}^{j}\eta_{25}^{(i)} < r - \sup_{\mathbb{T}^d_{\xi_0}}|v^{(0)}(\theta) - y_0|_\infty \ .
\end{aligned}$$

2.7. Iso-energetic KAM Theorem

We are now ready to formulate an iso-energetic KAM theorem based upon the above analysis. For convenience, we formulate the theorem in a self-contained way (repeating, therefore, some of the definitions given in the previous sections).

THEOREM 2.7.1. *Let $d \ge 2$ and let H be a scalar, real-analytic function on*[2.21] $\mathbb{T}^d_{\bar{\xi}} \times D^d_r(y_0)$ *for some $y_0 \in \mathbb{R}^d$. Let $E_{p,q}$ be positive numbers such that*

$$\begin{aligned}
&\|H_x\|_{\bar{\xi},r} \le E_{1,0} \ , && \|H_y\|_{\bar{\xi},r} \le E_{0,1} \ , \\
&\|H_{xy}\|_{\bar{\xi},r} \le E_{1,1} \ , && \|H_{yy}\|_{\bar{\xi},r} \le E_{0,2} \ , \\
&\|H_{xxy}\|_{\bar{\xi},r} \le E_{2,1} \ , && \|H_{xyy}\|_{\bar{\xi},r} \le E_{1,2} \ , \\
&\|H_{xxx}\|_{\bar{\xi},r} \le E_{3,0} \ , && \|H_{yyy}\|_{\bar{\xi},r} \le E_{0,3} \ .
\end{aligned} \tag{2.7.1}$$

Let (u, v, ω) be an approximate KAM torus satisfying

$$\begin{aligned}
\omega + D_\omega u - H_y(\theta + u, v) &= f \ , \\
D_\omega v + H_x(\theta + u, v) &= g \ , \\
u(0) &= 0 \ , \\
H(0, v(0)) - E &= h \ ,
\end{aligned}$$

where: $D_\omega := \omega\cdot\partial_\theta := \sum_{j=1}^d \omega_j \frac{\partial}{\partial\theta_j}$; ω is (γ,τ)-Diophantine, i.e.,

$$|\omega\cdot n| := \Big|\sum_{j=1}^d \omega_j n_j\Big| \ge \frac{\gamma}{|n|^\tau} \ , \qquad \forall\, n \in \mathbb{Z}^d\backslash\{0\} \ , \tag{2.7.2}$$

[2.20] Just recall that $v^{(j)} = v^{(0)} + \sum_{i=0}^{j-1} w^{(i)}$ and $\|w^{(i)}\|_{\xi_i} \le \eta_{25}^{(i)}$; see (2.3.16) and (2.6.20).

[2.21] $\mathbb{T}^d_\xi := \{y \in \mathbb{C}^d : |\operatorname{Im} y| \le \xi, \ \operatorname{Re} y_i \text{ defined mod } 2\pi\}$;
$D^d_r(y_0) := \{y \in \mathbb{C}^d : |y_i - y_{0i}| \le r, \ \forall\, i\}$.

for given numbers $\gamma > 0$, $\tau \ge d-1$; u and v vector-valued functions, which are real-analytic on $\mathbb{T}^d_\xi$ for some $0 < \xi \le \bar\xi$ and are assumed to satisfy

$$\sup_{\mathbb{T}^d_\xi} |\operatorname{Im} u| \le \bar\xi - \xi \,, \qquad \hat r := \sup_{\mathbb{T}^d_\xi} |v(\theta) - y_0|_\infty < r \,. \tag{2.7.3}$$

Fix $\rho > 0$. Assume that the $(d \times d)$-matrix

$$\mathcal{M}(\theta) := I + u_\theta$$

is invertible on $\mathbb{T}^d$ and that so is also the $((d+1)\times(d+1))$-matrix defined as

$$\mathcal{A} := \begin{pmatrix} \langle \mathcal{T} \rangle & -\langle \chi \rangle \\ \chi(0) & 0 \end{pmatrix} \,, \tag{2.7.4}$$

where:

$$\mathcal{T} := \mathcal{M}^{-1} H_{yy}(\theta + u, v) \mathcal{M}^{-T} \,,$$
$$\chi(\theta) = \chi(\theta;\rho) := \frac{1}{\rho} \mathcal{M}^{-1} H_y(\theta + u(\theta), v(\theta))$$

and $\langle\cdot\rangle$ denotes average over[2.22] *$\mathbb{T}^d$. Let F, G, $\overline{h}$, M, $\overline{M}$, $\overline{A}$, $\tilde V$, Ω be non-negative numbers such that*

$$\begin{aligned} &\|f\|_\xi \le F \,, \qquad \|g\|_\xi \le G \,, \qquad |h| \le \overline{h} \,, \\ &\|\mathcal{M}\|_\xi \le M \,, \qquad \|\mathcal{M}^{-1}\|_\xi \le \overline{M} \,, \qquad |\mathcal{A}^{-1}| \le \overline{A} \,, \\ &\|v_\theta\|_\xi \le \tilde V \,, \qquad |\omega| \le \Omega \,. \end{aligned} \tag{2.7.5}$$

Define the following weighted norms:

$$\begin{aligned}
E_1^* &:= \max\Big\{ E_{0,1} \,,\ \frac{E_{1,0}}{\rho} \Big\} \,, \\
E_2^* &:= \max\Big\{ E_{0,2} \,,\ \frac{E_{1,1}}{\rho} \Big\} \,, \\
E_3^* &:= \max\Big\{ E_{0,3} \,,\ \frac{E_{1,2}}{\rho} \,,\ \frac{E_{2,1}}{\rho^2} \,,\ \frac{E_{3,0}}{\rho^3} \Big\} \,, \\
E^* &:= \max\Big\{ E_1^* \,,\ E_2^* \rho \,,\ E_3^* \rho^2 \Big\} \,, \\
\Omega^* &:= \max\Big\{ \Omega \,, E^* \Big\} = \max\Big\{ \Omega \,,\ E_1^* \,,\ E_2^* \rho \,,\ E_3^* \rho^2 \Big\} \,, \\
\beta_0 &:= \max\Big\{ 1 \,,\ \frac{\tilde V}{\rho} \Big\} \,, \\
\beta_1 &:= \frac{\Omega^*}{\gamma} \,, \\
\alpha &:= \max\Big\{ 1 \,,\ \frac{\overline{A}\Omega^*}{\rho} \Big\} \,, \\
\mu &:= \max\Big\{ \frac{F}{\Omega} \,,\ \frac{G}{\Omega\,\rho} \,,\ \frac{\overline{h}}{\Omega\,\rho} \Big\} \,.
\end{aligned} \tag{2.7.6}$$

Fix

$$0 < \xi_\infty < \xi$$

[2.22] Note that the invertibility of the matrix (2.7.4) does not depend on the choice of $\rho > 0$, which should be interpreted as a free weight (having the physical dimension of the momenta y).

and let

$$\xi_* := \min\Big\{1 \ , \ \xi_\infty \ , \ \frac{\xi-\xi_\infty}{4}\Big\} \ . \tag{2.7.7}$$

Then, there exist constants $\hat c < c_*$ and c_{**} larger than one and depending only upon d, τ and κ such that the following holds. If μ is so small that

$$\begin{aligned} &\Big(c_* \ \overline{M}^{10} M^4 \ \xi_*^{-(4\tau+1)} \ \frac{\Omega^*}{\Omega}\alpha^2\beta_0^4\beta_1^4\Big) \ \mu \le 1 \ , \\ &\Big(c_{**} \ \overline{M}^5 M^2 \ \xi_*^{-2\tau} \ \alpha\beta_0^2\beta_1^2 \ \frac{\rho}{r-\hat r}\Big) \ \mu \le 1 \end{aligned} \tag{2.7.8}$$

then there exists a (unique) constant $\tilde a \in (-1,1)$ and (locally unique) functions $\tilde u$ and $\tilde v$, real-analytic on $\mathbb{T}^d_{\xi_\infty}$, satisfying[2.23]

$$\begin{aligned} \omega_{\tilde a} + D_{\tilde a}\tilde u - H_y(\theta+\tilde u,\tilde v) &= 0 \ , \\ D_{\tilde a}\tilde v + H_x(\theta+\tilde u,\tilde v) &= 0 \ , \\ \tilde u(0) &= 0 \ , \\ H(0,\tilde v(0)) &= E \ , \end{aligned}$$

and

$$\sup_{\mathbb{T}^d_{\xi_\infty}} |\operatorname{Im}\tilde u| < \bar\xi - \xi_\infty \ , \qquad \sup_{\mathbb{T}^d_{\xi_\infty}} |\tilde v(\theta) - y_0|_\infty < r \ .$$

Furthermore, $|\tilde a|$, $\|u-\tilde u\|_{\xi_\infty}$ and $\|v-\tilde v\|_{\xi_\infty}$ are small with μ, i.e.,

$$\begin{aligned} &\max\Big\{|\tilde a| \ , \ \|\tilde u - u\|_{\xi_\infty} \ , \ \|\tilde u_\theta - u_\theta\|_{\xi_\infty} \ , \ \rho^{-1}\|\tilde v - v\|_{\xi_\infty} \ , \ \rho^{-1}\|\tilde v_\theta - v_\theta\|_{\xi_\infty}\Big\} \\ &\le \Big(\hat c \ \overline{M}^5 M^2 \xi_*^{-(2\tau+1)} \ \alpha\beta_0^2\beta_1^2\Big) \ \mu \ . \end{aligned} \tag{2.7.9}$$

The proof of this theorem is based on Lemma 2.7.2 and Lemma 2.7.4 below.

LEMMA 2.7.2. *Fix $1 < \kappa \le 2$, and (using the notations of Theorem 2.7.1) let*

$$\xi_i := \xi_\infty + \frac{\xi-\xi_\infty}{2^i} \ , \qquad \delta_i := \frac{\xi-\xi_\infty}{2^{2+i}} \ .$$

Let

$$(u^{(0)}, v^{(0)}, \omega^{(0)}) := (u, v, \omega) \ ,$$

and, whenever it is defined for $i \ge 1$, let

$$\begin{aligned} &(u^{(i)}, v^{(i)}, \omega^{(i)}) := \mathcal{K}(u^{(i-1)}, v^{(i-1)}, \omega^{(i-1)}) \ , \\ &(\xi_i, \gamma_i, F_i, G_i, h_i, M_i, \overline{M}_i, U_i, V_i, \tilde V_i, \overline{A}_i) \\ &:= \widehat{\mathcal{K}}_{\delta_i,\sigma_{p,k}}(\xi_{i-1}, \gamma_{i-1}, F_{i-1}, G_{i-1}, h_{i-1}, M_{i-1}, \overline{M}_{i-1}, U_{i-1}, V_{i-1}, \tilde V_{i-1}, \overline{A}_{i-1}) \ , \end{aligned}$$

where $\mathcal{K}$ and $\widehat{\mathcal{K}}$ are the KAM maps defined in (2.4.4) and (2.6.31). Denote by $\eta_k^{(i)}$ the corresponding parameters appearing in the KAM estimates described in § 2.6 and let

$$\mu_i := \max\Big\{\frac{F_i}{\Omega} \ , \ \frac{G_i}{\Omega\,\rho} \ , \ \frac{\overline{h}_i}{\Omega\,\rho}\Big\} \ .$$

[2.23] As usual, $\omega_{\tilde a} := (1+\tilde a)\omega$, $D_{\tilde a} := D_{\omega_{\tilde a}} := \omega_{\tilde a}\cdot\partial_\theta$.

Let $j \geq 1$ and assume that, for $0 \leq i \leq j-1$, the following bounds hold:

$$\gamma_i^{-1} \leq \kappa\gamma^{-1} , \tag{2.7.10}$$

$$M_i \leq \kappa M , \tag{2.7.11}$$

$$\overline{M}_i \leq \kappa\overline{M} , \tag{2.7.12}$$

$$\overline{A}_i \leq \kappa\overline{A} , \tag{2.7.13}$$

$$\frac{\tilde{V}_i}{\rho} \leq \kappa\beta_0 , \tag{2.7.14}$$

$$\frac{\eta_5^{(i)}}{1-\eta_0^{(i)}} \leq (\kappa-1)\rho , \tag{2.7.15}$$

$$|v^{(i)}(0) - y_0|_\infty + \overline{M}_i\Big(\eta_2^{(i)} + \rho\eta_0^{(i)} + \frac{\eta_0^{(i)}\eta_5^{(i)}}{1-\eta_0^{(i)}}\Big) \leq r , \tag{2.7.16}$$

$$\overline{A}\eta_7^{(i)} \leq \frac{\kappa-1}{\kappa^2} , \tag{2.7.17}$$

$$\frac{(\eta_0^{(i)})^2}{1-\eta_0^{(i)}} E_{0,1}\overline{M}_i\eta_5^{(i)} \leq (\kappa-1)\frac{\Omega^*}{\rho}\frac{(\eta_9^{(i)})^2}{2} , \tag{2.7.18}$$

$$\frac{(\eta_0^{(i)})^2\eta_6^{(i)}}{1-\eta_0^{(i)}} \leq \kappa(\kappa-1)\frac{\Omega^*}{\rho^2}\frac{(\eta_9^{(i)})^2}{2} , \tag{2.7.19}$$

$$\left[\frac{2+\eta_0^{(i)}}{(1-\eta_0^{(i)})^2}\frac{\eta_0^{(i)}\eta_6^{(i)}}{\rho}\right] + \left[\frac{\Omega^*\eta_9^{(i)}}{\rho^3} + \eta_0^{(i)}\frac{2+\eta_0^{(i)}}{(1-\eta_0^{(i)})^2}\frac{\Omega^*}{\rho^2}\right]\overline{M}_i\eta_5^{(i)} \leq \frac{\Omega^*\eta_9^{(i)}\overline{M}_i}{\rho^2} , \tag{2.7.20}$$

$$\left[\frac{\Omega^*\eta_9^{(i)}}{\rho^3} + \eta_0^{(i)}\frac{2+\eta_0^{(i)}}{(1-\eta_0^{(i)})^2}\frac{\Omega^*}{\rho^2}\right]\overline{M}_i\eta_5^{(i)} \leq (\kappa^2-1)\frac{\Omega^*\eta_9^{(i)}\overline{M}_i}{\rho^2} , \tag{2.7.21}$$

$$(\eta_{10}^{(i)})^2 \leq (\kappa-1)\eta_8^{(i)} , \tag{2.7.22}$$

$$B\eta_{11}^{(i)} \leq \frac{\kappa-1}{\kappa^3} , \tag{2.7.23}$$

$$\eta_{13}^{(i)} \leq \frac{(\kappa-1)}{\kappa}\rho , \tag{2.7.24}$$

$$\frac{\eta_{13}^{(i)}/\rho}{1-(\eta_{13}^{(i)}/\rho)}\xi_*^{-(\tau+1)}\beta_1 2^{(\tau+1)i} \le \kappa - 1 \ , \tag{2.7.25}$$

$$\overline{M}\eta_{24}^{(i)} \le \frac{\kappa-1}{\kappa^2} \ , \tag{2.7.26}$$

$$E_{0,2}\eta_{28}^{(i)} \le (\kappa-1)\ 2E_{0,2}\overline{M}_i \ , \tag{2.7.27}$$

$$2\overline{M}_i\eta_{28}^{(i)} + (\eta_{28}^{(i)})^2 \le (\kappa-1)\overline{M}_i^2 \ , \tag{2.7.28}$$

$$\eta_{28}^{(i)} \le \kappa(\kappa-1)\overline{M} \ , \tag{2.7.29}$$

$$\overline{A}\ \sqrt{(\eta_{30}^{(i)})^2 + (\eta_{31}^{(i)})^2 + \eta_{30}^{(i)}\eta_{31}^{(i)}} \le \frac{\kappa-1}{\kappa^2} \ , \tag{2.7.30}$$

$$\eta_{23}^{(i)} \le 2\delta_i \ , \tag{2.7.31}$$

$$\eta_{25}^{(i)} \le r - \sup_{\mathbb{T}^d_{\xi_i}} |v^{(i)}(\theta) - y_0|_\infty \ . \tag{2.7.32}$$

Then, the functional KAM map $(u^{(i)}, v^{(i)}, \omega^{(i)}) = \mathcal{K}(u^{(i-1)}, v^{(i-1)}, \omega^{(i-1)})$ *is well defined for any* $1 \le i \le j$ *and, for any* $0 \le i \le j-1$*, one has*

$$\eta_k^{(i)} \ \le \ c_k\nu_k\chi_k B_k^i\ \mu_i \ , \tag{2.7.33}$$

$$q_k^{(i)} \ \le \ \bar{c}_k\bar{\nu}_k\bar{\chi}_k\bar{B}_k^i\ \mu_i^2 \ , \tag{2.7.34}$$

$$F_{i+1} \ \le \ \bar{c}_7\bar{\nu}_7\bar{\chi}_7\bar{B}_7^i\ \mu_i^2 \ , \tag{2.7.35}$$

$$G_{i+1} \ \le \ \bar{c}_8\bar{\nu}_8\bar{\chi}_8\bar{B}_8^i\ \mu_i^2 \ , \tag{2.7.36}$$

$$\bar{h}_{i+1} \ = \ 0 \ , \tag{2.7.37}$$

$$\mu_{i+1} \ \le \ \bar{c}_0\bar{\nu}_0\bar{\chi}_0\bar{B}_0^i\ \mu_i^2 \ , \tag{2.7.38}$$

where the constants c_k*'s,* ν_k*'s,* χ_k*'s,* B_k*'s,* $\bar{c}_k$*'s,* $\bar{\nu}_k$*'s,* $\bar{\chi}_k$*'s,* $\bar{B}_k$*'s are defined below; in* (2.7.33) *it is* $1 \le k \le 32$ *with* $k \ne 12, 16$ *(*$\eta_{12}^{(i)}$ *and* $\eta_{16}^{(i)}$ *do not appear any more), while in* (2.7.34) *it is* $1 \le k \le 6$*.*

Let

$$c := \max\left\{1\ ,\ \kappa\ \left(\frac{\tau}{\exp(1)}\right)^\tau\right\}\ , \qquad c^{(p)} := \max\left\{1\ ,\ \sqrt{d}\ \kappa^p\left(\frac{1+p\tau}{\exp(1)}\right)^{(1+p\tau)}\right\}\ .$$

Then:

$$c_1 := 2\kappa\ , \quad \nu_1 := M\ , \quad \chi_1 := \Omega\rho\beta_0\ , \quad B_1 := 1\ ,$$

$$c_2 := 2\kappa c\ , \quad \nu_2 := M\xi_*^{-\tau}\ , \quad \chi_2 := \rho\beta_0\beta_1\ , \quad B_2 := 1\ ,$$

$$c_3 := \kappa\ , \quad \nu_3 := \overline{M}\ , \quad \chi_3 := \Omega\ , \quad B_3 := 1\ ,$$

$$c_4 := \kappa^2 c_2\ , \quad \nu_4 := \overline{M}^2 M\xi_*^{-\tau}\ , \quad \chi_4 := \Omega^*\beta_0\beta_1\ , \quad B_4 := 1\ ,$$

$$c_5 := 2c^{(2)}\ , \quad \nu_5 := \xi_*^{-(2\tau+1)}\ , \quad \chi_5 := \rho\beta_1^2\ , \quad B_5 := 1\ ,$$

$$c_6 := \kappa^2 c_5\ , \quad \nu_6 := \overline{M}^2\xi_*^{-(2\tau+1)}\ , \quad \chi_6 := E_2^*\rho\beta_1^2\ , \quad B_6 := 1\ ,$$

$$c_7 := \max\left\{\sqrt{(\kappa^2 c_2)^2 + \kappa^3 c_2 c_5}\ , \sqrt{(c_3+c_6)^2 + \kappa^2 c_5^2 + \kappa^3 c_2 c_5}\right\}\ ,$$
$$\nu_7 := \overline{M}^2 M\xi_*^{-(2\tau+1)}\ , \quad \chi_7 := \frac{\Omega^*}{\rho}\beta_0\beta_1^2, \quad B_7 := 1$$

$$c_8 := \sqrt{(c_3+c_4)^2 + (1+\kappa c_2)^2}\ , \quad \nu_8 := \overline{M}^2 M\xi_*^{-\tau}\ ,$$
$$\chi_8 := \Omega^*\beta_0\beta_1\ , \quad B_8 := 1\ ,$$

$$c_0 := \kappa^3 c_8\ , \quad \nu_0 := \overline{M}^2 M\xi_*^{-\tau}\ , \quad \chi_0 := \alpha\beta_0\beta_1\ , \quad B_0 := 1\ ,$$

$$c_9 := \kappa^2 c_0\ , \quad \nu_9 := \overline{M}^3 M\xi_*^{-\tau}\ , \quad \chi_9 := \rho\chi_0\ , \quad B_9 := 1\ ,$$

$$c_{10} := \frac{\kappa c_9}{\sqrt{2}}\ , \quad \nu_{10} := \nu_9\ , \quad \chi_{10} := \sqrt{\Omega^*}\chi_0\ , \quad B_{10} := 1\ ,$$

$$c_{11} := \kappa^2 c_9\ , \quad \nu_{11} := \overline{M}^4 M\xi_*^{-\tau}\ , \quad \chi_{11} := \frac{\Omega^*}{\rho}\chi_0\ , \quad B_{11} := 1\ ,$$

$$c_{13} := \kappa^3 c_8\ , \quad \nu_{13} := \overline{M}^2 M\xi_*^{-\tau}\ , \quad \chi_{13} := \rho\chi_0\ , \quad B_{13} := 1\ ,$$

$$c_{14} := \kappa + c_{13}\ , \quad \nu_{14} := \overline{M}^2 M\xi_*^{-\tau}\ , \quad \chi_{14} := \Omega^*\chi_0\ , \quad B_{14} := 1\ ,$$

$$c_{15} := \kappa + c_{13}\ , \quad \nu_{15} := \overline{M}^2 M\xi_*^{-\tau}\ , \quad \chi_{15} := \Omega^*\rho\chi_0\ , \quad B_{15} := 1\ ,$$

$$c_{17} := \kappa^2 c_{13}\ , \quad \nu_{17} := \overline{M}^4 M\xi_*^{-\tau}\ , \quad \chi_{17} := \Omega^*\chi_0\ , \quad B_{17} := 1\ ,$$

$$c_{18} := cc_1\ , \quad \nu_{18} := M\xi_*^{-\tau}\ , \quad \chi_{18} := \rho\beta_0\beta_1\ , \quad B_{18} := 2^\tau\ ,$$

$$c_{19} := c_5\ , \quad \nu_{19} := \xi_*^{-(2\tau+1)}\ , \quad \chi_{19} := \rho\beta_1^2\ , \quad B_{19} := 2^{2\tau+1}\ ,$$

$$c_{20} := c_{18} + (\kappa - 1)c_{19} \ , \quad \nu_{20} := \nu_{18} \ , \quad \chi_{20} := \chi_{18} \ , \quad B_{20} := B_{18} \ ,$$

$$c_{21} := c_{17} + \kappa^2 c_{20} + \kappa c_{14} \ , \quad \nu_{21} := \overline{M}^4 M \xi_*^{-\tau} \ , \quad \chi_{21} := \Omega^* \chi_0 \ , \quad B_{21} := 2^\tau \ ,$$

$$c_{22} := \kappa c c_{21} \ , \quad \nu_{22} := \overline{M}^4 M \xi_*^{-2\tau} \ , \quad \chi_{22} := \beta_1 \chi_0 \ , \quad B_{22} := 2^\tau \ ,$$

$$c_{23} := \kappa(\kappa c c_{21} + c_{22}) \ , \quad \nu_{23} := \overline{M}^4 M^2 \xi_*^{-2\tau} \ , \quad \chi_{23} := \chi_{22} \ , \quad B_{23} := 2^{2\tau} \ ,$$

$$c_{24} := \Big(\frac{1}{2}\kappa^2 c^{(0)} c + \kappa c^{(1)}\Big) c_{21} + \frac{1}{2}\kappa c^{(0)} c_{22} \ ,$$
$$\nu_{24} := \overline{M}^4 M^2 \xi_*^{-(2\tau+1)} \ , \quad \chi_{24} := \chi_{22} \ , \quad B_{24} := 2^{2\tau+1} \ ,$$

$$c_{25} := \kappa(\kappa c_{23} + c_{13} + c_{20}),$$
$$\nu_{25} := \overline{M}^5 M^2 \xi_*^{-2\tau} \ , \quad \chi_{25} := \rho\beta_0\beta_1\chi_0 \ , \quad B_{25} := 2^{2\tau} \ ,$$

$$c_{26} := \frac{1}{2} c^{(0)} c_{25} + \kappa\Big(\frac{1}{2}\kappa c^{(0)} c_{23} + \kappa c_{24} + c^{(0)} c_{20}\Big) \ ,$$
$$\nu_{26} := \overline{M}^5 M^2 \xi_*^{-(2\tau+1)} \ , \quad \chi_{26} = \chi_{25} \ , \quad B_{26} := 2^{2\tau+1} \ ,$$

$$c_{27} := 2\kappa c(\kappa c^{(0)} c_{14} + \kappa c^{(0)} c_{15}) \ ,$$
$$\nu_{27} := \overline{M}^2 M^2 \xi_*^{-(2\tau+1)} \ , \quad \chi_{27} := \chi_{25} \ , \quad B_{27} := 2^{\tau+1} \ ,$$

$$c_{28} := \kappa^3 c_{24} \ , \quad \nu_{28} := \overline{M}^6 M^2 \xi_*^{-(2\tau+1)} \ , \quad \chi_{28} := \chi_{24} \ , \quad B_{28} := 2^{2\tau+1} \ ,$$

$$c_{29} := c_{23} + c_{25} \ , \quad \nu_{29} := \overline{M}^5 M^2 \xi_*^{-2\tau} \ , \quad \chi_{29} := E_3^* \rho\beta_0\beta_1\chi_0 \ , \quad B_{29} := 2^{2\tau} \ ,$$

$$c_{30} := 2\kappa^2 c_{28} + \kappa^3 c_{29} \ ,$$
$$\nu_{30} := \overline{M}^7 M^2 \xi_*^{-(2\tau+1)} \ , \quad \chi_{30} := \frac{E^*}{\rho}\beta_0\beta_1\chi_0 \ , \quad B_{30} := 2^{2\tau+1} \ ,$$

$$c_{31} := \kappa^2(c_{23} + c_{25}) + c_{28} \ ,$$
$$\nu_{31} := \overline{M}^6 M^2 \xi_*^{-(2\tau+1)} \ , \quad \chi_{31} := \chi_{30} \ , \quad B_{31} := B_{30} \ ,$$

$$c_{32} := \kappa^3 \sqrt{c_{30}^2 + c_{31}^2 + c_{30} c_{31}} \ ,$$
$$\nu_{32} := \overline{M}^7 M^2 \xi_*^{-(2\tau+1)} \ , \quad \chi_{32} := \overline{A}\alpha\beta_0\beta_1\chi_0 \ , \quad B_{32} := 2^{2\tau+1} \ ,$$

$$\bar{c}_1 := \frac{1}{2} c_{23}^2 + c_{23} c_{25} + \frac{1}{2} c_{25}^2 \ ,$$
$$\bar{\nu}_1 := \overline{M}^{10} M^4 \xi_*^{-4\tau} \ , \quad \bar{\chi}_1 := E_3^* \rho^2 \beta_0^2 \beta_1^2 \chi_0^2 \ , \quad \bar{B}_1 := 2^{4\tau} \ ,$$

$$
\begin{aligned}
&\bar{c}_2 := \frac{\kappa}{2} c^{(0)} c_{14} c_{23} \ , \\
&\qquad \bar{\nu}_2 := \overline{M}^7 M^3 \xi_*^{-(3\tau+1)} \ , \quad \bar{\chi}_2 := \Omega^* \beta_1 \chi_0^2 \ , \quad \bar{B}_2 := 2^{2\tau+1} \ , \\
&\bar{c}_3 := \kappa^2 c_{23} c_{27} \ , \\
&\qquad \bar{\nu}_3 := \overline{M}^8 M^4 \xi_*^{-(4\tau+1)} \ , \quad \bar{\chi}_3 := E_2^* \rho \beta_0 \beta_1^2 \chi_0^2 \ , \quad \bar{B}_3 := 2^{3\tau+1} \ , \\
&\bar{c}_4 := \bar{c}_1 \ , \quad \bar{\nu}_4 := \bar{\nu}_1 \ , \quad \bar{\chi}_4 := E_3^* \rho^3 \beta_0^2 \beta_1^2 \chi_0^2 \ , \quad \bar{B}_4 := \bar{B}_1 \ , \\
&\bar{c}_5 := \frac{\kappa}{2} c^{(0)} c_{15} c_{23} \ , \\
&\qquad \bar{\nu}_5 := \overline{M}^7 M^3 \xi_*^{-(3\tau+1)} \ , \quad \bar{\chi}_5 := \Omega^* \rho \beta_1 \chi_0^2 \ , \quad \bar{B}_5 := 2^{2\tau+1} \ , \\
&\bar{c}_6 := \frac{\kappa}{2} c^{(0)} c_{14} c_{25} \ , \\
&\qquad \bar{\nu}_6 := \overline{M}^8 M^3 \xi_*^{-(3\tau+1)} \ , \quad \bar{\chi}_6 := \Omega^* \rho \beta_0 \beta_1 \chi_0^2 \ , \quad \bar{B}_6 := 2^{2\tau+1} \ , \\
&\bar{c}_7 := \bar{c}_1 + \bar{c}_2 + \bar{c}_3 \ , \\
&\qquad \bar{\nu}_7 := \overline{M}^{10} M^4 \xi_*^{-(4\tau+1)} \ , \quad \bar{\chi}_7 := \Omega^* \beta_0^2 \beta_1^2 \chi_0^2 \ , \quad \bar{B}_7 := 2^{4\tau} \ , \\
&\bar{c}_8 := \bar{c}_4 + \bar{c}_5 + \bar{c}_6 + \kappa^2 (\bar{c}_2 + \bar{c}_3) \ , \\
&\qquad \bar{\nu}_8 := \overline{M}^{10} M^4 \xi_*^{-(4\tau+1)} \ , \quad \bar{\chi}_8 := \Omega^* \rho \beta_0^2 \beta_1^2 \chi_0^2 \ , \quad \bar{B}_8 := 2^{4\tau} \ , \\
&\bar{c}_0 := \max\{\bar{c}_7 \ , \ \bar{c}_8\} \ , \\
&\qquad \bar{\nu}_0 := \bar{\nu}_7 \ , \quad \bar{\chi}_0 := \frac{\Omega^*}{\Omega} \beta_0^2 \beta_1^2 \chi_0^2 = \frac{\Omega^*}{\Omega} \alpha^2 \beta_0^4 \beta_1^4 \ , \quad \bar{B}_0 := 2^{4\tau} \ .
\end{aligned}
\tag{2.7.39}
$$

REMARK 2.7.3. (i) The definition of ξ_* and the inductive assumptions

$$
\gamma^{(i)} \geq \kappa^{-1} \, \gamma^{(i-1)} \qquad (\forall \ 1 \leq i \leq j-1)
$$

imply immediately that $s_{p,k}(\xi_i, \omega^{(i)})$ and $s_{p,k}(\delta_i, \omega^{(i)})$ verify, for $0 \leq i \leq j-1$, the following bounds

$$
\begin{aligned}
s_{1,0}(\xi_i, \omega^{(i)}) &\leq c\gamma^{-1} \xi_i^{-\tau} \leq c\gamma^{-1} \xi_*^{-\tau} \\
&=: c \, (\Omega^*)^{-1} \beta_1 \xi_*^{-\tau} \ , \\
s_{p,1}(\xi_i, \omega^{(i)}) &\leq c^{(p)} \, \gamma^{-p} \, \xi_i^{-(p\tau+1)} \leq c^{(p)} \, \gamma^{-p} \, \xi_*^{-(p\tau+1)} \\
&=: c^{(p)} \, (\Omega^*)^{-p} \beta_1^p \, \xi_*^{-(p\tau+1)} \ ,
\end{aligned}
$$

$$\begin{aligned} s_{1,0}(\delta_i,\omega^{(i)}) &\le c\gamma^{-1}\delta_i^{-\tau} \le c\gamma^{-1}\xi_*^{-\tau}2^{\tau i}\\ &=: c(\Omega^*)^{-1}\beta_1\xi_*^{-\tau}2^{\tau i}\ , \end{aligned}$$

$$\begin{aligned} s_{p,1}(\delta_i,\omega^{(i)}) &\le c^{(p)}\,\gamma^{-p}\,\delta_i^{-(p\tau+1)} \le c^{(p)}\,\gamma^{-p}\,\xi_*^{-(p\tau+1)}2^{(p\tau+1)i}\\ &=: c^{(p)}\,(\Omega^*)^{-p}\beta_1^p\,\xi_*^{-(p\tau+1)}2^{(p\tau+1)i}\ . \end{aligned} \tag{2.7.40}$$

Clearly, for $i=0$, (2.7.40) holds without any further assumption.

(ii) Notice that

$$\frac{\Omega}{\gamma}\ge 1\ ,$$

as it follows from (2.7.2) by taking as n the versor $e^{(j)}$ (so that $|\omega_j|\ge\gamma$, $\forall j$). Thus $\beta_1\ge 1$. Furthermore,

$$\min\{M,\overline{M}\}\ge 1\ ;$$

in fact,

$$M\ge\|\mathcal{M}\|_\xi=\|\mathcal{M}^T\|_\xi\ge\|\mathcal{M}^T\|_0=\|\mathcal{M}\|_0\ ,$$

and

$$\mathcal{M}^Te^{(1)}=e^{(1)}+\frac{\partial u_1}{\partial\theta}\ .$$

Thus, if θ_0 is a critical point of the multi-periodic function $\theta\to u_1(\theta)$, we have that

$$\mathcal{M}^T(\theta_0)e^{(1)}=e^{(1)}\ , \tag{2.7.41}$$

and (recall the notation in § 2.5)

$$\begin{aligned} \|\mathcal{M}^T\|_0 &:= |N_0(\mathcal{M}^T)|=\sup_{c\in\mathbb{C}^d:|c|=1}\sqrt{\sum_i\Big|\sum_j\|\mathcal{M}_{ji}\|_0c_j\Big|^2}\\ &\ge \sqrt{\sum_i\Big|\sum_j|\mathcal{M}_{ji}(\theta_0)|e_j^{(1)}\Big|^2}=1\ . \end{aligned}$$

Clearly, since, by (2.7.41), one has

$$\mathcal{M}^{-T}(\theta_0)e^{(1)}=e^{(1)}\ ,$$

the same argument applies to $\mathcal{M}^{-1}$. We, therefore, have

$$\min\Big\{M,\overline{M},\frac{\Omega}{\gamma},\frac{\Omega^*}{\Omega},\beta_k,c_k,\bar c_k,\nu_k,\bar\nu_k,B_k,\bar B_k\Big\}\ge 1\ .$$

(iii) By (2.7.13), one sees that (2.7.17) implies

$$\overline{A}^{(i)}\eta_7^{(i)}\le\frac{\kappa-1}{\kappa}\qquad\Longleftrightarrow\qquad\frac{1}{1-\overline{A}^{(i)}\eta_7^{(i)}}\le\kappa\ ,$$

which, in turn, implies

$$B^{(i)}:=\frac{\overline{A}^{(i)}}{1-\overline{A}^{(i)}\eta_7^{(i)}}\le\kappa^2\overline{A}^{(i)}\ . \tag{2.7.42}$$

(iv) The twenty three inductive assumptions (2.7.10)÷(2.7.32) include the eight conditions (2.6.23) ÷ (2.6.30) plus fifteen conditions, which are used to *simplify* the bounds given in (2.7.33) (in the same fashion as (2.7.17) has been used in the previous point to simplify the estimate of $B^{(i)}$).

For later use we specify where, in the proof of Lemma 2.7.2, it is used each single assumption in[2.24] (2.7.10)÷(2.7.32):

the eight conditions (2.6.23)÷(2.6.30) correspond to or are implied by, respectively, (2.7.16), (2.7.22), (2.7.23), (2.7.24), (2.7.26), (2.7.30), (2.7.31) and (2.7.32);
(2.7.10)÷(2.7.14) are used systematically to derive (2.7.33) (for $i \geq 1$);
(2.7.33) for k =1, 2, 3, 4, 5, 6, 7, 8, 0, 18 and 19 *needs no other inductive assumptions in order to be derived* besides (2.7.10)÷(2.7.14); the estimates on $\eta_{12}^{(i)}$ and $\eta_{16}^{(i)}$ do not appear in the sequel;
(2.7.17) $\rightarrow$ (2.7.42);
(2.7.15) $\rightarrow$ (2.7.33) with $k = 9$;
(2.7.18) and (2.7.19) $\rightarrow$ (2.7.33) with $k = 10$;
(2.7.20) and (2.7.21) $\rightarrow$ (2.7.33) with $k = 11$;
(2.7.23) $\rightarrow$ (2.7.33) with $k = 13$;
(2.7.24) $\rightarrow$ (2.7.33) with $k = 14$;
(2.7.23) $\rightarrow$ (2.7.33) with $k = 15$;
(2.7.23) $\rightarrow$ (2.7.33) with $k = 17$;
(2.7.25) $\rightarrow$ (2.7.33) with $k = 20$;
(2.7.23)÷(2.7.25) $\rightarrow$ (2.7.33) with $k = 21, 22, 23, 24, 25, 26$;
(2.7.23) and (2.7.24) $\rightarrow$ (2.7.33) with $k = 27$;
(2.7.23)÷(2.7.26) $\rightarrow$ (2.7.33) with $k = 28$;
(2.7.23)÷(2.7.25) $\rightarrow$ (2.7.33) with $k = 29$;
(2.7.23)÷(2.7.28) $\rightarrow$ (2.7.33) with $k = 30$;
(2.7.23)÷(2.7.26) and (2.7.29) $\rightarrow$ (2.7.33) with $k = 31$;
(2.7.23)÷(2.7.30) $\rightarrow$ (2.7.33) with $k = 32$.

(v) In view of the definitions of $\bar{\nu}_0$, $\bar{\chi}_0$ and χ_0, the first condition in (2.7.8) may be rewritten as

$$\mu \leq \frac{1}{c_* \bar{\nu}_0 \bar{\chi}_0} \ . \tag{2.7.43}$$

(vi) A quadratic relation (among positive numbers) of the form

$$\mu_{i+1} \leq a b^i \mu_i^2 \ , \qquad i \geq 0 \ , \tag{2.7.44}$$

is equivalent to the relation

$$\hat{\mu}_{i+1} \leq \hat{\mu}_i^2 \ , \qquad \hat{\mu}_i := a b^{i+1} \mu_i \ . \tag{2.7.45}$$

Iterating (2.7.45), one gets

$$\hat{\mu}_i \leq \hat{\mu}_0^{2^i} \ , \qquad i \geq 0 \ ,$$

which corresponds to

$$\mu_i \leq \frac{(a b \mu_0)^{2^i}}{a b^{i+1}} \ .$$

In particular, if

$$\hat{\mu}_0 = a b \mu_0 \leq 1 \ , \tag{2.7.46}$$

then $\hat{\mu}_i \leq 1$ for all i and

$$\hat{\mu}_{i+1} \leq \hat{\mu}_i^2 \leq \hat{\mu}_i \ ,$$

which yields $\hat{\mu}_i \leq \hat{\mu}_0$, i.e.,

$$\mu_i \leq \frac{\mu_0}{b^i} \leq \mu_0 \ . \tag{2.7.47}$$

[2.24] We use the notation "(a) $\rightarrow$ (b)" meaning "(a) is used in the derivation of (b)".

Indeed, (2.7.38) has the form (2.7.44) with $a := \bar{c}_0\bar{\nu}_0\bar{\chi}_0$ and $b := \bar{B}_0$. Therefore, *assuming that*

$$c_* \geq \bar{B}_0\bar{c}_0 = 2^{4\tau}\bar{c}_0 \tag{2.7.48}$$

(as we shall do later: see (2.7.58) and (2.7.57) below), we see that (2.7.43) implies (2.7.46), so that (2.7.47), i.e.,

$$\mu_i \leq \frac{\mu_0}{\bar{B}_0^i} = \frac{\mu}{2^{4\tau i}} , \tag{2.7.49}$$

holds under the hypotheses of Lemma 2.7.2 (and (2.7.48)) for $i \leq j$.

(vii) For later use, we note that

$$\begin{aligned}
&\bar{\nu}_0 \geq \nu_k , \qquad \forall\, k , \\
&\frac{B_k}{\bar{B}_0} \leq \frac{2^{2\tau+1}}{2^{4\tau}} \leq \frac{1}{2} , \qquad \forall\, k , \\
&\bar{\chi}_0 \geq \max\left\{\beta_1\chi_0, \chi_{24}, \chi_{28}, \overline{A}^{-1}\chi_{32}, \frac{\chi_{26}}{\rho}, \frac{\chi_5}{\rho}, \frac{\chi_{25}}{\rho}, \overline{A}\chi_7, \overline{A}E_2^*\chi_0\right\} .
\end{aligned} \tag{2.7.50}$$

PROOF. (of Lemma 2.7.2) The estimating technique, based on a systematic use of points (i)÷(iv) of Remark 2.7.3, is rather straightforward and goes along the following lines:

a) use (2.7.10)÷(2.7.14) and (2.7.42) to get rid of the i-dependence in the parameters M_i, $\overline{M}_i$, $\overline{A}_i$, $\tilde{V}_i$ and B_i;
c) define the upper bounds $\sigma_{p,k} \geq s_{p,k}$ as the right hand sides of (2.7.40);
d) in each expression factor out the terms (all of which are greater or equal than one) M, $\overline{M}$, ξ_*^{-1} with the maximal power with which they appear: this term gives the ν_k or the $\bar{\nu}_k$;
e) in each expression factor out the term 2^i with the maximal power: this term gives the B_k or the $\bar{B}_k$;
f) in each expression factor out the maximal term involving γ, Ω, $E_{p,q}$, ρ, $\tilde{V}$, $\overline{A}$, F_i, G_i, $\bar{h}_i$ expressing it in term of[2.25] β_0, β_1, E_p^*, α and μ_i: this term will give the $\chi_k\mu_i$ or $\bar{\chi}_k\mu_i^2$;
g) the expression that remains after the above operations gives the numerical constant c_k or $\bar{c}_k$.

To illustrate the above strategy we start by performing, in details, the first few estimates. For clarity, we denote by $(2.7.33)_k$ the estimate (2.7.33). Let us begin with $(2.7.33)_1$. By (2.7.14), (2.7.11), the definition of μ_i and the fact that β_0 and M are greater or equal than one, we find

$$\begin{aligned}
\eta_1^{(i)} &:= \tilde{V}_iF_i + M_iG_i \leq \kappa(\rho\beta_0F_i + MG_i) \leq \kappa(\rho\beta_0\Omega + M\Omega\rho)\mu_i \\
&\leq (2\kappa)M\ (\Omega\rho\beta_0)\ \mu_i =: c_1\nu_1\chi_1B_1^i\mu_i ,
\end{aligned}$$

proving $(2.7.33)_1$. Estimate $(2.7.33)_2$ is gotten similarly using (2.7.40). Estimate $(2.7.33)_3$ is obvious.

Since, by the definitions (2.7.6), it is

$$\rho E_{0,2} \leq \rho E_2^* \leq \Omega^* , \tag{2.7.51}$$

[2.25] In doing this operation there is, in general, some freedom and we shall try, in the following, to discuss it with some care.

one finds, using (2.7.12), $(2.7.33)_2$ and (2.7.51),

$$\begin{aligned} \eta_4^{(i)} &:= \overline{M}_i^2 E_{0,2}\eta_2^{(i)} \le \kappa^2 \overline{M}^2 E_{0,2}\eta_2^{(i)} \\ &\le (\kappa^2 c_2)(\overline{M}^2 M\xi_*^{-\tau})(\rho E_{0,2}\beta_0\beta_1)\mu_i \le (\kappa^2 c_2)(\overline{M}^2 M\xi_*^{-\tau})(\Omega^*\beta_0\beta_1)\mu_i \\ &=: c_4\nu_4\chi_4 B_4^i\mu_i\ . \end{aligned}$$

Observe that

$$\max\Big\{\frac{E_{1,0}}{\rho}, E_{0,1}\Big\} = E_1^*\ , \qquad \max\Big\{\frac{E_1^*}{\Omega^*}, \frac{\Omega}{\Omega^*}\Big\} \le 1\ . \tag{2.7.52}$$

Then, by (2.7.40) and (2.7.52), one finds

$$\begin{aligned} \eta_5^{(i)} &:= \sigma_{2,1}(\xi_i, \omega^{(i)})(E_{1,0}F_i + E_{0,1}G_i) \\ &\le \sigma_{2,1}(\xi_i, \omega^{(i)})\Omega\rho\Big(\frac{E_{1,0}}{\rho}\frac{F_i}{\Omega} + E_{0,1}\frac{G_i}{\Omega\rho}\Big) \\ &\le (2c^{(2)})\xi_*^{-(2\tau+1)}(\rho\beta_1^2)\frac{\Omega}{\Omega^*}\frac{E_1^*}{\Omega^*}\mu_i \\ &\le (2c^{(2)})\xi_*^{-(2\tau+1)}(\rho\beta_1^2)\mu_i \\ &=: c_5\nu_5\chi_5 B_5^i\mu_i\ . \end{aligned}$$

Estimate $(2.7.33)_6$ follows at once from

$$E_{0,2}\chi_5 := E_{0,2}\rho\beta_1^2 \le E_2^*\rho\beta_1^2 \le \Omega^*\beta_1^2 =: \chi_6\ .$$

Let us turn to $(2.7.33)_7$. Recalling (2.6.11) and (2.6.10), one finds that

$$\eta_7 \le c_7\nu_7\ \frac{1}{\rho}\max\Bigg\{E_{0,2}\chi_2, \sqrt{E_{0,2}\chi_2\frac{E_{0,1}\chi_5}{\rho}}, \chi_3, \chi_6, \frac{E_{0,1}\chi_5}{\rho}\Bigg\}\ \mu_i\ ,$$

and[2.26]

$$\begin{aligned} &\frac{1}{\rho}\max\Bigg\{E_{0,2}\chi_2, \sqrt{E_{0,2}\chi_2\frac{E_{0,1}\chi_5}{\rho}}, \chi_3, \chi_6, \frac{E_{0,1}\chi_5}{\rho}\Bigg\} \\ &\qquad \le \frac{1}{\rho}\max\Big\{E_{0,2}\chi_2, \chi_3, \chi_6, \frac{E_{0,1}\chi_5}{\rho}\Big\} \\ &\qquad \le \frac{1}{\rho}\max\Big\{E_2^*\rho\beta_0\beta_1, \Omega, E_2^*\rho\beta_1^2, E_{0,1}\beta_1^2\Big\} \\ &\qquad \le \frac{\Omega^*}{\rho}\beta_0\beta_1^2 =: \chi_7\ , \end{aligned}$$

which proves $(2.7.33)_7$.

Estimate $(2.7.33)_8$ follows observing that

$$\begin{aligned} \max\Big\{\chi_3, \chi_4, \Omega, \frac{E_{0,1}}{\rho}\chi_2\Big\} &\le \max\Big\{\Omega, E_2^*\rho\beta_0\beta_1, E_{0,1}\beta_0\beta_1\Big\} \\ &\le (\beta_0\beta_1)\max\Big\{\Omega, E_2^*\rho, E_{0,1}\Big\} \\ &\le \Omega^*\beta_0\beta_1 =: \chi_8\ . \end{aligned}$$

Estimate $(2.7.33)_0$ follows easily recalling (2.7.42) and the definition of α.

[2.26] For any $a, b \ge 0$, $\sqrt{ab} \le \max\{a, b\}$.

By (2.7.15) one gets

$$\overline{M}\rho\eta_0^{(i)} \le \eta_9^{(i)} \le \kappa^2\overline{M}\rho\eta_0^{(i)} \ . \tag{2.7.53}$$

The second estimate in (2.7.53) together with $(2.7.33)_0$ implies at once $(2.7.33)_9$. Using (2.7.18) and (2.7.19), one finds that[2.27]

$$(\eta_{10}^{(i)})^2 \le \kappa^2 \frac{\Omega^*}{\rho^2} \frac{(\eta_9^{(i)})^2}{2} \ , \tag{2.7.54}$$

which, by $(2.7.33)_9$, implies immediately $(2.7.33)_{10}$.
By (2.7.20) and (2.7.21), one finds that[2.28]

$$\eta_{11}^{(i)} \le \kappa \frac{\Omega^*}{\rho^2} \eta_9^{(i)} \overline{M}_i \ ,$$

from which $(2.7.33)_{11}$ follows.
As mentioned above, the explicit estimate on η_{12} is never needed.
As pointed out in (iii) of Remark 2.7.3, (2.7.13) and (2.7.17) imply (2.7.42), i.e.,

$$B^{(i)} \le \kappa^2 \overline{A} \ . \tag{2.7.55}$$

By (2.7.23) and (2.7.55), one finds

$$\eta_{13}^{(i)} \le \kappa^3 \overline{A} \eta_8^{(i)} \ ,$$

from which $(2.7.33)_{13}$ follows.
Condition (2.7.24) implies that

$$\eta_{13}^{(i)} \le \rho(\kappa - 1) \ , \tag{2.7.56}$$

which, in turn, implies that

$$\begin{aligned} \eta_{14}^{(i)} &\le \kappa F_i + \eta_{13}^{(i)} \frac{E_{0,1}}{\rho} \le (\kappa\Omega + c_{13}\nu_{13}\chi_0 E_{0,1})\ \mu_i \\ &\le c_{14}\nu_{14}\chi_{14}\mu_i \ . \end{aligned}$$

The estimate $(2.7.33)_{15}$ is completely analogous (using, again, (2.7.56)).
As mentioned above, the explicit estimate on η_{16} is never needed.
The estimate $(2.7.33)_{17}$ is obvious (since $E_{0,2}\rho \le \Omega^*$).
Also the estimates $(2.7.33)_{18}$ and $(2.7.33)_{19}$ are completely straightforward.
Condition (2.7.25) implies that

$$\eta_{20}^{(i)} \le \eta_{18}^{(i)} + \frac{\kappa - 1}{\xi_*^{-(\tau+1)}\beta_1 2^{(\tau+1)i}} \eta_{19}^{(i)} \ ,$$

which yields easily $(2.7.33)_{20}$.
Estimates $(2.7.33)_{21} \div (2.7.33)_{29}$ follow, now, easily along the above lines (using (2.7.24) for $(2.7.33)_{22}$, $(2.7.33)_{23}$ and $(2.7.33)_{24}$ and (2.7.26) for $(2.7.33)_{28}$).
By (2.7.27) and (2.7.28) one finds

$$\eta_{30}^{(i)} \le 2\kappa E_{0,2}\overline{M}_i\eta_{28}^{(i)} + \kappa\overline{M}_i^2\eta_{29}^{(i)} \ ,$$

[2.27] If a, b, c are non negative numbers such that $c \le (\kappa-1)b$ and $a \le \kappa(\kappa-1)b$ (which is less than $\kappa\sqrt{\kappa^2-1}b$), then $\sqrt{a^2+(b+c)^2} \le \kappa^2 b$.

[2.28] If a, b, c are non negative numbers such that $b+c+d \le a$, then $b^2+(c+d)^2 \le a^2$ so that $\max\{a^2, b^2+(c+d)^2\} = a^2$; furthermore, if $c+d \le (\kappa^2-1)a$ then $\sqrt{a^2+a(c+d)} \le \kappa a$.

which yields immediately $(2.7.33)_{30}$.
By (2.7.29) one finds

$$\eta_{31}^{(i)} \le \frac{1}{\rho}\Big[\kappa^2 \overline{M}(E_{1,1}\eta_{23}^{(i)} + E_{0,2}\eta_{25}^{(i)}) + E_{0,1}\eta_{28}^{(i)}\Big] ,$$

which yields immediately $(2.7.33)_{31}$.
Finally, by (2.7.30) we get

$$\eta_{32}^{(i)} \le \overline{A}^2 \kappa^3 \sqrt{(\eta_{30}^{(i)})^2 + (\eta_{31}^{(i)})^2 + \eta_{30}^{(i)}\eta_{31}^{(i)}} ,$$

from which $(2.7.33)_{32}$ follows easily.
At this point, the estimate (2.7.34) (with $1 \le k \le 6$), as well as (2.7.35)÷(2.7.38), are immediately checked. □

LEMMA 2.7.4. *Under the hypotheses of Theorem 2.7.1, the bounds listed in (2.7.10)÷(2.7.32) and in (2.7.33)÷(2.7.38) hold for any $i \ge 0$. In fact, the constants c_* and c_{**} in (2.7.8) and $\hat{c}$ in (2.7.9) can be taken as follows. Let*[2.29]

$$\hat{c}_1 := \kappa^2 c_{13}\frac{\bar{B}_0}{\bar{B}_0 - 1} , \quad \hat{c}_2 := 2c_{24} , \quad \hat{c}_3 := 2c_{28} , \quad \hat{c}_4 := 2c_{32} ,$$

$$\hat{c}_5 := 2c_{26} , \quad \hat{c}_6 := \kappa c_0 , \quad \hat{c}_7 := \kappa c_5 ,$$

$$\hat{c}_9 := \kappa^2 c_7 , \quad \hat{c}_{10} := 2\kappa^2 c_5 , \quad \hat{c}_{11} := 2c_6 , \quad \hat{c}_{12} := 3\kappa^5 c_5 ,$$

$$\hat{c}_{13} := \frac{1}{2}\kappa^9 c_0 , \quad \hat{c}_{14} := \kappa^3 c_{11} , \quad \hat{c}_{15} := \kappa c_{13} , \quad \hat{c}_{16} := \hat{c}_{15} ,$$

$$\hat{c}_{17} := \kappa^2 c_{24} , \quad \hat{c}_{18} := 2\kappa c_{28} , \quad \hat{c}_{19} := \frac{c_{32}}{\kappa^3} .$$

and

(2.7.57) $$\hat{c}_0 := \max_{1 \le i \le 19, i \ne 8} \hat{c}_i , \quad \tilde{c}_0 := \max\{2^{4\tau}\bar{c}_0, 3\kappa^4 c_6\} .$$

Then, one can take

$$\hat{c} := 2\max\{c_{13}, c_{23}, c_{24}, c_{25}, c_{26}\} , \quad c_* := \max\left\{\tilde{c}_0, \frac{\hat{c}_0}{\kappa - 1}\right\} ,$$

$$c_{**} := \frac{4}{3}c_{25} + \kappa c_2 + \kappa^2 c_0 .$$

(2.7.58)

REMARK 2.7.5. (i) The proof of this lemma is by induction: first one shows that (2.7.8) with c_* and c_{**} given in (2.7.58) imply (2.7.10)÷(2.7.32) with $i = 0$; by Lemma 2.7.2 it then follows that (2.7.33)÷(2.7.38) hold for $i = 0$; then one fixes $j \ge 1$ and assumes that (2.7.10)÷(2.7.32) hold for $0 \le i \le j-1$ and proves that, thanks to (2.7.8), they hold also for $i = j$; again, by Lemma 2.7.2, it will then follow that also (2.7.33)÷(2.7.38) hold for any $i = j$ completing the inductive argument.
(ii) As above, we shall attach an index $_0$, $_i$ or $_j$ to the inequalities (2.7.10)÷(2.7.38) to specify the index with which such inequalities are considered.

[2.29]The constants c_k and $\bar{B}_0$ are listed in (2.7.39). Note that $\hat{c}_8$ does not exist.

(iii) Note that in the proof of $(2.7.33)_i$ for k =1, 2, 3, 4, 5, 6, 7, 8, 0, 18 and 19, the assumptions $(2.7.10)_i \div (2.7.32)_i$ *are not needed* (compare also (iv), Remark 2.7.3). In particular, $(2.7.33)_0$ *with the above listed k's hold without any further assumption.*

(iv) In the first part of the following proof, i.e., in the proof of $(2.7.10)_0 \div (2.7.32)_0$, there will appear powers of κ, sometimes, in excess: the reason for this will be plain in the second part of the proof (compare also with, e.g., footnotes 2.31 and 2.32 below).

(v) To give some explicit values: in the case $d = 2$, $\tau = 1$, $\kappa = 1.01$, one can take

$$\hat{c} = 111.7 \;, \qquad c_* = 38528.282 \;, \qquad c_{**} = 49.088 \;.$$

(vi) The above constants allow to get an idea of what is the range of applicability of the KAM Theorem 2.7.1 without extra work. To be concrete, let us consider a simple ("the simplest") non-integrable system (to which much attention has been devoted; compare, e.g., [**51**], [**50**], [**23**], [**37**]), namely the forced pendulum

$$H(x_1, x_2, y_1, y_2; \varepsilon) := \frac{y_1^2}{2} + y_2 + \varepsilon\Big(\cos x_1 + \cos(x_1 - x_2)\Big) \;. \tag{2.7.59}$$

One is interested in the analytic continuation of the unperturbed torus $\mathbb{T}^2 \times \{\omega_g, 1\}$ where ω_g denotes the golden mean $(\sqrt{5} - 1)/2$. To apply Theorem 2.7.1 to the trivial approximant solution

$$(u, v, \omega) := \Big((0,0), (\omega_g, 1), (\omega_g, 1)\Big) \tag{2.7.60}$$

we make the following choices[2.30]

$$\begin{aligned} &\bar{\xi} = \xi = 3.65172 \;, \quad \xi_\infty = \frac{\xi}{5} \;, \\ &r = 0.1 \;, \quad y_0 = (\omega_g, 1) \;, \quad \hat{r} = 0 \;, \quad \rho = 1.17557 \;. \end{aligned} \tag{2.7.61}$$

With such choices, one finds immediately:

$$\begin{aligned} &\Omega^* = \Omega = \rho = E^* = 1.17558 \;, \quad \overline{A} = 1.978 \;, \quad \tilde{V} = 0 \;, \\ &\beta_0 = 1 \;, \quad \beta_1 = \frac{\Omega}{\gamma} \;, \quad \alpha = \overline{A} \;, \quad F = 0 \;, \\ &G = |\varepsilon| \sqrt{\Big(\cosh\xi + \cosh\sqrt{2}\xi\Big)^2 + (\cosh\sqrt{2}\xi)^2} \le |\varepsilon| \; 137.99982 \;, \\ &\overline{h} = 2|\varepsilon| \;, \quad M = \overline{M} = 1 \;, \quad \mu = \frac{G}{\Omega\rho} \;; \end{aligned} \tag{2.7.62}$$

and both KAM conditions (2.7.8) are implied by the first condition, which reads

$$|\varepsilon| \le 1.5 \cdot 10^{-10} \;. \tag{2.7.63}$$

PROOF. (of Lemma 2.7.4) First, we show how (2.7.8) imply $(2.7.10)_0 \div (2.7.32)_0$. *In the following estimates, besides* (2.7.8) *and* (2.7.50), *we shall make use point (iii) of Remark 2.7.5.* We recall also (iv) and (v) of Remark 2.7.3 and, in particular, (2.7.43), which is equivalent to the first condition in (2.7.8).

$(2.7.10)_0 \div (2.7.13)_0$ are obvious since $\kappa > 1$.

[2.30] It is an easy exercise to check that this choices are essentially optimal for the trivial initial approximate solution (2.7.60); the Diophantine constants for the vector $(\omega_g, 1)$ are $\tau = 1$ and $\gamma = 2/(\sqrt{5} + 3)$ (see, e.g., [**24**]).

$(2.7.14)_0$ follows from the definition of β_0.

Now we prove that

$$\frac{1}{1-\eta_0^{(0)}} \le \kappa \,. \tag{2.7.64}$$

In fact, by $(2.7.33)_0$ with $k = 0$, (2.7.43) and (2.7.50),

$$\begin{aligned} \eta_0^{(0)} &\le c_0\nu_0\chi_0\mu \le c_0\nu_0\chi_0 \frac{1}{c_*\bar{\nu}_0\bar{\chi}_0} \\ &\le \frac{c_0}{c_*} \le \frac{\kappa-1}{\kappa} \,, \end{aligned} \tag{2.7.65}$$

where the last inequality holds since (by definition of c_*)

$$c_* \ge \frac{1}{\kappa-1}\,\hat{c}_6 := \frac{\kappa}{\kappa-1} c_0 \,.$$

Inequality (2.7.65) is equivalent to (2.7.64).

$(2.7.15)_0$: by (2.7.64), $(2.7.33)_0$ with $k = 5$, (2.7.43) and (2.7.50),

$$\begin{aligned} \frac{\eta_5^{(0)}}{1-\eta_0^{(0)}} &\le \kappa\eta_5^{(0)} \le \kappa c_5\nu_5\chi_5\mu \le \kappa c_5\nu_5\chi_5 \frac{1}{c_*\bar{\nu}_0\bar{\chi}_0} \\ &\le \frac{\kappa c_5}{c_*}\,\rho =: \frac{\hat{c}_7}{c_*}\,\rho \le (\kappa-1)\rho \,, \end{aligned} \tag{2.7.66}$$

where the last inequality holds by definition of c_*.
The bound $(2.7.33)_0$ *with* $k = 9$ *then follows* (compare (iv), Remark 2.7.3).

$(2.7.16)_0$: by definition of $\hat{r}$ (see (2.7.3)), by $(2.7.15)_0$, $(2.7.33)_0$ with $k = 0$ and $k = 2$, the definitions of χ_2 and χ_0 (which imply that $\chi_2 \le \rho\chi_0$), the definition of c_{**}, the second condition in (2.7.8),

$$\begin{aligned} |v(0) - y_0|_\infty + \overline{M}\Big(\eta_2^{(0)} + \rho\eta_0^{(0)} + \frac{\eta_0^{(0)}\eta_5^{(0)}}{1-\eta_0^{(0)}}\Big) &\le \hat{r} + \overline{M}\Big(\eta_2^{(0)} + \kappa\rho\eta_0^{(0)}\Big) \\ &\le \hat{r} + \overline{M}\Big(c_2\nu_2\chi_2 + \kappa\rho c_0\nu_0\chi_0\Big)\mu \\ &\le \hat{r} + (c_2 + \kappa c_0)\rho\overline{M}^2 M\xi_*^{-\tau}\chi_0\mu \\ &\le \hat{r} + c_{**}\rho\overline{M}^2 M\xi_*^{-\tau}\chi_0\mu \\ &< \hat{r} + (r - \hat{r}) = r \,. \end{aligned}$$

$(2.7.17)_0$: by $(2.7.33)_0$ with $k = 7$, (2.7.43) and (2.7.50)

$$\begin{aligned} \overline{A}\eta_7^{(0)} &\le \frac{\overline{A}c_7\nu_7\chi_7}{c_*\bar{\nu}_0\bar{\chi}_0} \le \frac{c_7}{c_*} \\ &=: \frac{\hat{c}_9}{\kappa^2 c_*} \le \frac{\kappa-1}{\kappa^2} \,. \end{aligned}$$

The bound $(2.7.42)_0$ *then follows* (compare (iii), Remark 2.7.3).

$(2.7.18)_0$: from the definition of $\eta_9^{(i)}$, (2.6.14), it follows that

$$\eta_9^{(i)} \ge \overline{M}_i\rho\eta_0^{(i)} \ge \overline{M}\rho\eta_0^{(i)} \,; \tag{2.7.67}$$

thus (2.7.18) is implied by

$$\frac{\eta_5^{(i)}}{1-\eta_0^{(i)}}\ E_{0,1} \le \frac{\kappa-1}{2\kappa}\ \Omega^* \rho\ .$$

Setting $i=0$, using the first line in (2.7.66), (2.7.50), we get

$$\begin{aligned}\frac{\eta_5^{(i)}}{1-\eta_0^{(i)}}\ E_{0,1} &\le \kappa c_5 \nu_5 \chi_5 \frac{1}{c_* \bar{\nu}_0 \bar{\chi}_0} E_{0,1} \le \frac{\kappa c_5}{c_*}\ \rho \Omega^* \\ &=: \frac{\hat{c}_{10}}{2\kappa}\ \frac{1}{c_*}\ \rho\Omega^* \le \frac{\kappa-1}{2\kappa}\Omega^* \rho\ ,\end{aligned}$$

where last inequality holds by definition of c_*.

$(2.7.19)_0$: by (2.7.67) and (2.7.64) one sees that (2.7.19) is implied by

$$\eta_6^{(0)} \le \frac{1}{2}(\kappa-1)\Omega^*\ . \tag{2.7.68}$$

Using $(2.7.33)_0$ with $k=6$, we see that (2.7.68) holds because of (2.7.43), (2.7.50) and the fact that

$$c_* \ge \frac{1}{\kappa-1}\hat{c}_{11} := \frac{1}{\kappa-1}(2c_6)\ .$$

The bound $(2.7.33)_0$ *with* $k=10$ *then follows* (compare (iv), Remark 2.7.3).

$(2.7.20)_0$ and $(2.7.21)_0$: it is immediate to check that $(2.7.20)_i$ and $(2.7.21)_i$ are implied by the following three inequalities:

$$\frac{2+\eta_0^{(i)}}{(1-\eta_0^{(i)})^2}\ \frac{\eta_0^{(i)}\eta_6^{(i)}}{\rho} \le \frac{1}{\kappa^2}\ \frac{\Omega^* \eta_9^{(i)} \overline{M}_i}{\rho^2}\ , \tag{2.7.69}$$

$$\frac{\Omega^* \eta_9^{(i)} \overline{M}_i \eta_5^{(i)}}{\rho^3} \le \frac{\kappa-1}{\kappa}\ \frac{\Omega^* \eta_9^{(i)} \overline{M}_i}{\rho^2}\ , \tag{2.7.70}$$

$$\frac{2+\eta_0^{(i)}}{(1-\eta_0^{(i)})^2}\ \frac{\Omega^*}{\rho^2}\ \overline{M}_i\ \eta_0^{(i)}\eta_5^{(i)} \le \frac{\kappa-1}{\kappa^2}\ \frac{\Omega^* \eta_9^{(i)} \overline{M}_i}{\rho^2}\ . \tag{2.7.71}$$

Notice that (2.7.64) implies that

$$\frac{2+\eta_0^{(0)}}{(1-\eta_0^{(0)})^2} \le \kappa^2\Big(3-\frac{1}{\kappa}\Big) < 3\kappa^2\ .$$

Thus, by (2.7.53) one has that (2.7.69)÷(2.7.71) (with $i=0$) are implied by

$$\begin{aligned}\eta_6^{(0)} &\le \frac{1}{3\kappa^4}\Omega^*\ , \\ \eta_5^{(0)} &\le \rho\frac{\kappa-1}{\kappa^2}\ , \\ \frac{1}{\rho}\ \eta_5^{(0)} &\le \frac{\kappa-1}{3\kappa^5}\ .\end{aligned}$$

Proceeding as above one easily checks that these inequalities hold since (recalling the definition of $\hat{c}_{12}$ and c_*)

$$c_* \ge \max\Big\{3\kappa^4 c_6, \frac{1}{\kappa-1}\hat{c}_{12}\Big\}\ .$$

The bound $(2.7.33)_0$ *with* $k=11$ *then follows* (compare (iv), Remark 2.7.3).

$(2.7.22)_0$: by (2.7.54) and (2.7.67) (with $i=0$) one finds[2.31]

$$(\eta_{10}^{(0)})^2 \le \frac{\kappa^6}{2}\Omega^* \,\overline{M}^2 (\eta_0^{(0)})^2 \; . \tag{2.7.72}$$

Recalling the definition of $\eta_0^{(i)}$ (see (2.6.12)), one finds

$$\eta_8^{(i)} = \frac{\eta_0^{(i)}\rho}{\kappa B^{(i)}} \; ,$$

so that, for $i=0$, recalling that $B \le \kappa^2 \overline{A}$,

$$\eta_8^{(0)} \ge \frac{\eta_0^{(0)}\rho}{\kappa^3 \overline{A}} \; ,$$

and $(2.7.22)_0$, in view also of the definition of α (in (2.7.6)), is seen to be implied by[2.32]

$$\frac{\kappa^9}{2}\;\alpha\;\overline{M}^2\;\eta_0^{(0)}\;\frac{1}{\kappa-1} \le 1 \; . \tag{2.7.73}$$

But, by definition of $\eta_0^{(0)}$, by (2.7.43), the relations

$$\bar\nu_0 \ge \overline{M}^2 \nu_0 \; , \qquad \bar\chi_0 \ge \alpha\chi_0 \; , \tag{2.7.74}$$

(which follow by definition), the definitions of $\hat c_{13}$ and of c_*, one finds

$$\begin{aligned}
\frac{\kappa^9}{2}\;\alpha\;\overline{M}^2\;\eta_0^{(0)}\;\frac{1}{\kappa-1} \;&\le\; \frac{\kappa^9}{2}\;\alpha\;\overline{M}^2\;\frac{c_0\nu_0\chi_0}{c_*\bar\nu_0\bar\chi_0}\;\frac{1}{\kappa-1}\\
&\le\; \frac{\kappa^9}{2}\;\frac{c_0}{c_*}\frac{1}{\kappa-1} =: \frac{\hat c_{13}}{c_*}\;\frac{1}{\kappa-1}\\
&\le\; 1 \; ,
\end{aligned}$$

showing the validity of (2.7.73) and, hence, of $(2.7.22)_0$.

$(2.7.23)_0$: observing that, by definition of χ_{11}, the definition of α and (2.7.74),

$$\overline{A}\chi_{11} = \overline{A}\frac{\Omega^*}{\rho}\;\chi_0 \le \alpha\;\chi_0 \le \bar\chi_0$$

one sees that $(2.7.23)_0$ is immediately implied by the definition of $\hat c_{14}$ and the definition of c_*. *The bounds* $(2.7.33)_0$ *with* $k = 13, 15$ *and* 17 *then follow* (compare (iv), Remark 2.7.3).

$(2.7.24)_0$: follows from the definitions of χ_{13}, $\hat c_{15}$ and c_*. *The bound* $(2.7.33)_0$ *with* $k = 14$ *then follows* (compare (iv), Remark 2.7.3).

$(2.7.25)_0$: using $(2.7.24)_i$ one sees that $(2.7.25)_i$ is implied by

$$\frac{\eta_{13}^{(i)}}{\rho}\;\kappa\;\xi_*^{-(\tau+1)}\;\beta_1 2^{(\tau+1)i} \le \kappa - 1 \; ,$$

which, for $i=0$, holds because

$$\frac{\chi_{13}}{\rho}\;\beta_1 \le \bar\chi_0 \; , \qquad \overline{M}^2 M \xi_*^{-(2\tau+1)} \le \bar\nu_0 \; ,$$

[2.31] The bound (2.7.72) holds, in fact, with κ^2 in place of κ^6. However, later (in the case $i>0$) the power κ^6 will appear; compare also point (iv) of Remark 2.7.5.

[2.32] Here would be enough to require the bound (2.7.73) with κ^5 replacing κ^9; however, since later we shall need the stronger condition with κ^9, we require it already here.

and because $c_* \geq (\kappa-1)^{-1}\hat{c}_{16} := (\kappa-1)^{-1}\kappa c_{13}$. *The bounds* $(2.7.33)_0$ *with* $21 \leq k \leq 27$ *and* $k = 29$ *then follow* (compare (iv), Remark 2.7.3).

$(2.7.26)_0$: it holds because $c_* \geq (\kappa-1)^{-1}\hat{c}_{17} := (\kappa-1)^{-1}\kappa^2 c_{24}$. *The bound* $(2.7.33)_0$ *with* $k = 28$ *then follows* (compare (iv), Remark 2.7.3).

$(2.7.27)_0$, $(2.7.28)_0$ and $(2.7.29)_0$: first notice that $(2.7.27)_i$ is implied by $(2.7.28)_i$, which, in turn, is implied by[2.33]

$$\eta_{28}^{(i)} \leq \frac{\kappa-1}{2\kappa}\,\overline{M}\ . \tag{2.7.75}$$

Such relation implies also $(2.7.29)_i$. *Thus* (2.7.75) *implies* $(2.7.27)_i$, $(2.7.28)_i$ *and* $(2.7.29)_i$. For $i = 0$, (2.7.75) is immediately seen to hold since $c_* \geq (\kappa-1)^{-1}\hat{c}_{18} := (\kappa-1)^{-1}2\kappa c_{28}$. *The bounds* $(2.7.33)_0$ *with* $k = 30$ *and* 31 *then follow* (compare (iv), Remark 2.7.3).

$(2.7.30)_0$: using the bounds $(2.7.33)_0$ with $k = 30$ and 31, the definitions of c_{32}, ν_{32} and χ_{32}, one finds

$$\overline{A}\sqrt{(\eta_{30}^{(0)})^2 + (\eta_{31}^{(0)})^2 + \eta_{30}^{(0)}\eta_{31}^{(0)}} \leq \frac{c_{32}}{\kappa^3}\ \nu_{32}\ \frac{\chi_{32}}{\overline{A}}\mu \leq \kappa - 1\ ,$$

since

$$\nu_{32} \leq \bar{\nu}_0\ , \qquad \alpha\beta_0\beta_1\chi_0 \leq \bar{\chi}_0\ ,$$

and $c_* \geq (\kappa-1)\hat{c}_{19} = (\kappa-1)^{-1}\ c_{32}/\kappa^3$.
The bound $(2.7.33)_0$ *with* $k = 32$ *then follows* (compare (iv), Remark 2.7.3).

$(2.7.31)_0$: the condition (2.7.31) is equivalent to

$$\frac{c_{23}}{2}\ \xi_*^{-1}\nu_{23}\ \chi_{23}\ 2^i B_{23}^i \mu_i \leq 1\ ,$$

which, for $i = 0$, is satisfied since[2.34] $c_* \geq c_{23}$.

$(2.7.32)_0$: by $(2.7.33)_0$ with $k = 25$, the definition of ν_{25}, χ_{25}, χ_0, c_{**} and the second condition in (2.7.8), we find

$$\begin{aligned}
\sup_{\mathbb{T}^d_\xi}|v - y_0|_\infty + \eta_{25}^{(0)} &\leq \hat{r} + c_{25}\nu_{25}\chi_{25}\mu \\
&\leq \hat{r} + c_{25}\overline{M}^5 M^2\ \xi_*^{-2\tau}\alpha\rho\ \beta_0^2\beta_1^2\ \mu \\
&\leq \hat{r} + c_{**}\overline{M}^5 M^2\ \xi_*^{-2\tau}\alpha\rho\ \beta_0^2\beta_1^2\ \mu \\
&\leq \hat{r} + r - \hat{r} = r\ .
\end{aligned}$$

This finishes the proof of $(2.7.10)_0 \div (2.7.32)_0$. Thus, by Lemma 2.7.2, $(2.7.34)_0 \div (2.7.38)_0$ follow.

Fix, now, $j \geq 1$ *and assume, by induction, that* $(2.7.10)_i \div (2.7.32)_i$ *hold for any* $0 \leq i \leq j-1$. By Lemma 2.7.2, $(2.7.33)_i \div (2.7.38)_i$ hold for any $0 \leq i \leq j-1$. If we show that $(2.7.10)_j \div (2.7.32)_j$ hold, then (again by Lemma 2.7.2) the proof of Lemma 2.7.4 will be complete.

2.33 $\dfrac{1}{\kappa} + \dfrac{\kappa-1}{4\kappa^2} < 1$.

2.34 In fact, by (2.7.58), (2.7.57) and the definitions in (2.7.39) one sees that (recall that $\tau \geq 1$)

$$c_* \geq \bar{c}_0 \geq 16\ \bar{c}_0 \geq 16\ \bar{c}_7 \geq 16\ \bar{c}_3 \geq 16\ c_{23}\ .$$

$(2.7.10)_j$: we shall use the following elementary inequality

$$(1-x) \geq \exp(-tx) \,, \qquad \forall\, t > 1 \,, \quad \forall\, 0 \leq x \leq 1 - \frac{1}{t} \,.$$

Thus, in view of $(2.7.24)_i$ (which implies that $\eta_{13}^{(i)}/\rho \leq 1 - 1/\kappa$) and in view of the definition of γ_i, we find

$$\begin{aligned} \gamma_j &:= \Big(1 - \frac{\eta_{13}^{(j-1)}}{\rho}\Big)\gamma_{j-1} \geq \gamma_{j-1} \exp\Big(- \kappa \frac{\eta_{13}^{(j-1)}}{\rho}\Big) \\ &\geq \gamma \exp\Big(- \frac{\kappa}{\rho} \sum_{i=0}^{j-1} \eta_{13}^{(i)}\Big) \\ &\geq \gamma\Big(1 - \frac{\kappa}{\rho}\sum_{i=0}^{j-1} \eta_{13}^{(i)}\Big) \,, \end{aligned} \tag{2.7.76}$$

(we used also the trivial bound $\exp(-x) \geq 1 - x$). Let us now prove that

$$\frac{1}{\rho} \sum_{i=0}^{j-1} \eta_{13}^{(i)} \leq \frac{\kappa - 1}{\kappa^2} \,. \tag{2.7.77}$$

In fact, by $(2.7.33)_i$ with $i \leq j-1$ (which hold because of the inductive assumption) and by (2.7.43), we get

$$\begin{aligned} \frac{\kappa^2}{\kappa - 1}\frac{1}{\rho}\sum_{i=0}^{j-1} \eta_{13}^{(i)} &\leq \frac{\kappa^2}{\kappa - 1} c_{13}\nu_{13}\chi_0 \sum_{i=0}^{j-1} \mu_i \\ &\leq \frac{\kappa^2}{\kappa - 1} c_{13}\nu_{13}\chi_0 \sum_{i=0}^{j-1} \frac{1}{c_* \bar\nu_0 \bar\chi_0 \bar B_0^i} \\ &= \Big(\frac{\kappa^2}{\kappa - 1}\frac{c_{13}}{c_*}\frac{\bar B_0}{\bar B_0 - 1}\Big)\frac{\nu_{13}}{\bar\nu_0}\frac{\chi_0}{\bar\chi_0} \\ &\leq 1 \,, \end{aligned}$$

where the last inequality holds because $\nu_{13} \leq \bar\nu_0$, $\chi_0 \leq \bar\chi_0$ and

$$c_* \geq \frac{1}{\kappa - 1}\, \hat c_1 := \frac{1}{\kappa - 1}\, \Big(\kappa^2 c_{13} \frac{\bar B_0}{\bar B_0 - 1}\Big) \,.$$

By (2.7.76) and (2.7.77), $(2.7.10)_j$ follows.

$(2.7.11)_j$: from now on we shall use systematically $(2.7.33)_i$ with $i \leq j - 1$ and

$$\frac{B_k}{\bar B_0} \leq \frac{1}{2} \,.$$

Recalling the definition of $M_{i+1} := M_i + \eta_{24}^{(i)}$, we find

$$\begin{aligned} M_j &\le M + \sum_{i=0}^{j-1} \eta_{24}^{(i)} \\ &\le M + c_{24}\nu_{24}\chi_{24} \sum_{i=0}^{j-1} B_{24}^i \mu_i \\ &\le M + \frac{c_{24}\nu_{24}\chi_{24}}{c_* \bar{\nu}_0 \bar{\chi}_0} \sum_{i=0}^{j-1} \left(\frac{B_{24}}{\bar{B}_0}\right)^i \\ &\le M + \frac{c_{24}\nu_{24}\chi_{24}}{c_* \bar{\nu}_0 \bar{\chi}_0} \sum_{i=0}^{j-1} \frac{1}{2^i} \\ &\le M + \frac{2c_{24}\nu_{24}\chi_{24}}{c_* \bar{\nu}_0 \bar{\chi}_0} \\ &\le M + \frac{2c_{24}}{c_*} \\ &\le \kappa M \ , \end{aligned}$$

where, besides (2.7.50), we have used the fact that $c_* \ge \hat{c}_2/(\kappa - 1) := 2c_{24}/(\kappa - 1)$.

$(2.7.12)_j$, $(2.7.13)_j$ and $(2.7.14)_j$: recall that

$$\overline{M}_{i+1} := \overline{M}_i + \eta_{28}^{(i)}, \qquad \overline{A}_{i+1} := \overline{A}_i + \eta_{32}^{(i)}, \qquad \tilde{V}_{i+1} := \tilde{V}_i + \eta_{26}^{(i)} \ .$$

Thus, replacing $\eta_{24}^{(i)}$ with, respectively, $\eta_{28}^{(i)}$, $\eta_{32}^{(i)}$ and $\eta_{26}^{(i)}$ in the above estimate on M_j one sees that $(2.7.12)_j$, $(2.7.13)_j$ and $(2.7.14)_j$ hold since $c_* \ge \hat{c}_k/(\kappa - 1)$ and $\hat{c}_3 := 2c_{28}$, $\hat{c}_4 := 2c_{32}$, $\hat{c}_5 := 2c_{26}$.

In exactly the same way we proved (2.7.64), one sees that, since $c_* \ge \hat{c}_6/(\kappa - 1)$,

$$\frac{1}{1 - \eta_0^{(i)}} \le \kappa \ , \qquad \forall\, i \le j \ . \tag{2.7.78}$$

$(2.7.15)_j$: in view of (2.7.78), one sees that the argument used to check $(2.7.15)_0$ extends immediately to $j > 0$.

$(2.7.16)_j$: by $(2.7.11)_j$ (which has been already proved), by $(2.7.15)_j$, by definition of $\hat{r}$ (see (2.7.3)), $(2.7.33)_i$ with $k = 25, 2, 0$ and $i \le j - 1$, (2.7.49) (which, by the inductive assumptions and Lemma 2.7.2, holds for $i \le j$), the definition of c_{**} and the second condition in (2.7.8), we find

$$
\begin{aligned}
|v^{(j)}(0) - y_0|_\infty &+ \overline{M}_j\Big(\eta_2^{(j)} + \rho\eta_0^{(j)} + \frac{\eta_0^{(j)}\eta_5^{(j)}}{1-\eta_0^{(j)}}\Big)\\
&\le |v(0) - y_0|_\infty + \sum_{i=0}^{j-1} |w^{(i)}(0)| + \kappa\overline{M}\Big(\eta_2^{(j)} + \kappa\rho\eta_0^{(j)}\Big)\\
&\le \hat{r} + \sum_{i=0}^{j-1} \eta_{25}^{(i)} + \kappa\overline{M}\Big(\eta_2^{(j)} + \kappa\rho\eta_0^{(j)}\Big)\\
&\le \hat{r} + c_{25}\nu_{25}\chi_{25}\sum_{i=0}^{j-1} B_{25}^i\mu_i + \kappa\overline{M}\Big(c_2\nu_2\chi_2 + \kappa\rho c_0\nu_0\chi_0\Big)\mu_j\\
&\le \hat{r} + \Big(c_{25}\nu_{25}\frac{\chi_{25}}{\rho}\sum_{i=0}^{j-1}\frac{1}{4^i} + \kappa c_2\overline{M}\nu_2\frac{\chi_2}{\rho} + \kappa^2 c_0\overline{M}\nu_0\chi_0\Big)\rho\mu\\
&\le \hat{r} + \Big(\frac{4}{3}c_{25} + \kappa c_2 + \kappa^2 c_0\Big)\ \nu_{25}\chi_{25}\ \mu\\
&\le \hat{r} + \rho\Big(c_{**}\ \overline{M}^5 M^2\ \xi_*^{-2\tau}\ \alpha\beta_0^2\beta_1^2\ \Big)\ \mu\\
&\le \hat{r} + (r - \hat{r}) = r\ .
\end{aligned}
$$

Using (2.7.49) (and playing some attention to the use of the inductive assumptions together with their consequences), one checks easily that *the argument used to derive* $(2.7.17)_0 \div (2.7.31)_0$ *may be used to derive also* $(2.7.17)_j \div (2.7.31)_j$; the proof of $(2.7.32)_j$ is analogous to that of $(2.7.16)_j$.

Finally, we prove (2.7.9). From the definitions given (compare, in particular, (2.7.58)), one sees immediately that[2.35] $\hat{c} < c_*$. Furthermore, recalling § 2.4, from what we have proved until now (in particular, that, thanks to (2.7.8), the KAM map is well defined for all $i \ge 0$) we have that

$$
\begin{aligned}
&\|\tilde{u} - u\|_{\xi_\infty} \le \sum_{i=0}^{\infty} \eta_{23}^{(i)}\ , \qquad \|\tilde{u}_\theta - u_\theta\|_{\xi_\infty} \le \sum_{i=0}^{\infty} \eta_{24}^{(i)}\ ,\\
&\|\tilde{v} - v\|_{\xi_\infty} \le \sum_{i=0}^{\infty} \eta_{25}^{(i)}\ , \qquad \|\tilde{v}_\theta - v_\theta\|_{\xi_\infty} \le \sum_{i=0}^{\infty} \eta_{26}^{(i)}\ ;
\end{aligned}
\tag{2.7.79}
$$

and, since

$$
\tilde{a} := \Big[\prod_{i=0}^{\infty}\Big(1 + a_i\Big)\Big] - 1\ ,
$$

we have also that

$$
|\tilde{a}| \le \exp\Big(\frac{1}{\rho}\sum_{i=0}^{\infty} \eta_{13}^{(i)}\Big) - 1\ ,
\tag{2.7.80}
$$

[2.35] The argument is similar to that used in the footnote 2.34.

From (2.7.79), (2.7.33) (with $k = 23, 24, 25, 26$), the definition of $\hat{c}$, (2.7.49), we find that

$$\begin{aligned}
&\max\Big\{\|\tilde{u} - u\|_{\xi_\infty}\ ,\ \|\tilde{u}_\theta - u_\theta\|_{\xi_\infty}\ ,\ \rho^{-1}\|\tilde{v} - v\|_{\xi_\infty}\ ,\ \rho^{-1}\|\tilde{v}_\theta - v_\theta\|_{\xi_\infty}\Big\}\\
&\le \frac{\hat{c}}{2}\ \overline{M}^5 M^2 \xi_*^{-(2\tau+1)} \alpha\beta_0^2\beta_1^2\ \ \mu \sum_{i=0}^\infty \Big(\frac{2^{2\tau+1}}{\bar{B}_0}\Big)^i\\
&\le \frac{\hat{c}}{2}\ \overline{M}^5 M^2 \xi_*^{-(2\tau+1)} \alpha\beta_0^2\beta_1^2\ \ \mu \sum_{i=0}^\infty \frac{1}{2^i}\\
&= \Big(\hat{c}\ \overline{M}^5 M^2 \xi_*^{-(2\tau+1)}\ \alpha\beta_0^2\beta_1^2\Big)\ \mu\ .
\end{aligned}$$

Next, observe that by (2.7.58), (2.7.57), (2.7.39) and (2.7.8) it follows that

$$c_{13} \le \frac{\hat{c}}{2}\ ,\qquad c_{13} \le \bar{c}_0 \le \frac{c_*}{16}\ ,\qquad \rho^{-1} c_{13}\nu_{13}\chi_{13}\mu \le \frac{1}{16}\ . \tag{2.7.81}$$

Thus,

$$\frac{1}{\rho}\sum_{i=0}^\infty \eta_{13}^{(i)} \le c_{13}\nu_{13}\chi_0\mu \sum_{i=0}^\infty \frac{1}{2^{4i}} = \frac{16}{15}\ c_{13}\nu_{13}\chi_0\mu\ ,$$

so that, by (2.7.80) and (2.7.81), one finds[2.36]

$$\begin{aligned}
|\tilde{a}|\ &\le\ \frac{16}{15}\ c_{13}\nu_{13}\chi_0\mu\ \Big(1 + \frac{16}{15}\ c_{13}\nu_{13}\chi_0\mu\Big)\\
&\le\ \frac{16}{15}\ c_{13}\nu_{13}\chi_0\mu\ \Big(1 + \frac{1}{15}\Big)\\
&=\ \Big(\frac{16}{15}\Big)^2\ \ c_{13}\nu_{13}\chi_0\mu\\
&<\ 2c_{13}\nu_{13}\chi_0\mu\\
&\le\ \Big(\hat{c}\ \overline{M}^5 M^2 \xi_*^{-(2\tau+1)}\ \alpha\beta_0^2\beta_1^2\Big)\ \mu\ .
\end{aligned}$$

completing the proof of (2.7.9). □

2.8. Iso-energetic Lindstedt series

In this section, we discuss shortly (convergent) power series expansions for KAM iso-energetic tori in terms of a smallness parameter. Such series (in the non-fixed energy case) are known as *Lindstedt series*. Here, we follow, essentially, Moser's original (indirect) argument ([**112**]).

2.8.1. Analytic dependence upon parameters. Roughly speaking, if the system depends analytically and uniformly on a set of parameters, then so do the KAM tori obtained via Theorem 2.7.1.

More precisely, let $\mathcal{E}$ be the closure of an open, connected, bounded set of parameters in $\mathbb{C}^m$ (with non-empty intersection with $\mathbb{R}^m$) and assume that, in Theorem 2.7.1,

$$\begin{aligned}
&H = H(x,y;\varepsilon)\ ,\quad E = E(\varepsilon)\ ,\quad \omega := \Big(1 + a(\varepsilon)\Big)\omega_0\ ,\\
&u = u(\theta;\varepsilon)\ ,\quad v = v(\theta;\varepsilon)\ ,
\end{aligned}\tag{2.8.1}$$

[2.36] Observe that $\exp(x) - 1 \le x(1+x)$ for any $0 \le x \le 1$.

are real-analytic (and bounded) also in the parameters $\varepsilon = (\varepsilon_1, ..., \varepsilon_m) \in \mathcal{E}$; thus, f, g and h will also depend analytically upon ε.

Let us, now, indicate the modifications that are needed in order to extend Theorem 2.7.1 to the present parameter-dependent case.

Assume that ω_0 is (γ_0, τ)-Diophantine and let

$$\gamma := (1 - \sup_{\mathcal{E}} |a|)\gamma_0 > 0 ,$$

so that (2.7.2) holds also in the present case *uniformly in* $\varepsilon \in \mathcal{E}$. Next, replace systematically the norms $\|\cdot\|_\xi$ and $\|\cdot\|_{\xi,r}$ with, respectively,

$$\|\cdot\|_{\xi,\mathcal{E}} := \sup_{\mathcal{E}} \|\cdot\|_\xi , \qquad \|\cdot\|_{\xi,r,\mathcal{E}} := \sup_{\mathcal{E}} \|\cdot\|_{\xi,r} .$$

Thus, for example, the $E_{p,q}$'s are upper bounds on $\|H.\|_{\xi,r,\mathcal{E}}$; (2.7.3) reads, now,

$$\sup_{\mathbb{T}^d_\xi,\mathcal{E}} |\operatorname{Im} u| \le \bar{\xi} - \xi , \qquad \hat{r} := \sup_{\mathbb{T}^d_\xi,\mathcal{E}} |v(\theta;\varepsilon) - y_0|_\infty < r ;$$

Ω in (2.7.5) is, now, an upper bound on $\sup_{\mathcal{E}} |\omega|$, which we shall take to be

$$\Omega \ge (1 + \sup_{\mathcal{E}} |a|)|\omega_0| .$$

Then: *if* (2.7.8) *holds, there exists a (locally unique) real-analytic solution* $(\tilde{u}, \tilde{v}) = (\tilde{u}(\theta;\varepsilon), \tilde{v}(\theta;\varepsilon))$ *of the system*

$$\begin{aligned} \omega_{\tilde{a}} + D_{\tilde{a}}\tilde{u} - H_y(\theta + \tilde{u}, \tilde{v}; \varepsilon) &= 0 , \\ D_{\tilde{a}}\tilde{v} + H_x(\theta + \tilde{u}, \tilde{v}; \varepsilon) &= 0 , \\ \tilde{u}(0;\varepsilon) &= 0 , \\ H(0, \tilde{v}(0); \varepsilon) &= E(\varepsilon) , \end{aligned} \tag{2.8.2}$$

where $\omega_{\tilde{a}} := (1 + \tilde{a})\omega$, $D_{\tilde{a}} := D_{\omega_{\tilde{a}}} := \omega_{\tilde{a}} \cdot \partial_\theta$; $\tilde{u}$ *and* $\tilde{v}$ *are real-analytic on* $\mathbb{T}^d_{\xi_\infty} \times \mathcal{E}$, $\tilde{a}$ *is real-analytic on* $\mathcal{E}$. *Furthermore,* $|\tilde{a}|$, $\|u - \tilde{u}\|$ *and* $\|v - \tilde{v}\|$ *are small with* μ:

$$\begin{aligned} &\max\Big\{ |\tilde{a}|_{\mathcal{E}} \ , \ \|\tilde{u} - u\|_{\xi_\infty,\mathcal{E}} \ , \ \|\tilde{u}_\theta - u_\theta\|_{\xi_\infty,\mathcal{E}} \ , \ \rho^{-1}\|\tilde{v} - v\|_{\xi_\infty,\mathcal{E}} \ , \\ &\rho^{-1}\|\tilde{v}_\theta - v_\theta\|_{\xi_\infty,\mathcal{E}} \Big\} \le \left(\hat{c}\, \overline{M}^5 M^2 \xi_*^{-(2\tau+1)} \ \alpha\beta_0^2\beta_1^2 \right) \mu . \end{aligned} \tag{2.8.3}$$

The proof of this statement rests upon the complex extension of Lemma 2.3.7 and on Weierstrass theorem (on the analyticity of uniform limits of analytic functions). The complex version of Lemma 2.3.7 (whose proof is basically identical to the real case and is omitted) reads as follows.

LEMMA 2.8.1. *Let* $\Omega := \{c \in \mathbb{C}^d : |c| \le \rho_1\} \times \{\hat{a} \in \mathbb{C} : |\hat{a}| \le \rho_2\}$ *for some* $\rho_1, \rho_2 > 0$; *let* $\alpha := (c, \hat{a})$ *and let* $(\alpha, \varepsilon) \in \Omega \times \mathcal{E} \subset \mathbb{C}^{d+1} \times \mathbb{C} \to F(\alpha;\varepsilon)$ *be a real-analytic function with values in* $\mathbb{C}^{d+1}$. *Assume that the Jacobian* F_α *is invertible for all* $(\alpha, \varepsilon) \in \Omega \times \mathcal{E}$. *Let* $F_\alpha(0;\varepsilon)^{-1} =: \mathcal{B} := \mathcal{B}(\varepsilon)$ *and assume that*

$$\begin{aligned} &\sup_{\Omega\times\mathcal{E}} |\alpha - \mathcal{B}(\varepsilon)F(\alpha;\varepsilon)| \le \min\{\rho_1, \rho_2\} , \\ &\sup_{\Omega\times\mathcal{E}} |I - \mathcal{B}(\varepsilon)\ F_\alpha(\alpha;\varepsilon)| < 1 . \end{aligned}$$

Then, there exists a unique real-analytic function $\varepsilon \in \mathcal{E} \to \alpha_0(\varepsilon) \in \Omega$ *such that*

$$F(\alpha_0(\varepsilon);\varepsilon) \equiv 0 .$$

Moreover, one has

$$\sup_{(\alpha,\varepsilon)\in\Omega\times\mathcal{E}} |(F_\alpha(\alpha;\varepsilon))^{-1}| \le |\mathcal{B}|_{\mathcal{E}} \left(1-\sup_{\Omega\times\mathcal{E}}|I-\mathcal{B}F_\alpha|\right)^{-1}$$

and

$$|\alpha_0|_{\mathcal{E}} \le |\mathcal{B}|_{\mathcal{E}} \left(1-\sup_{\Omega\times\mathcal{E}}|I-\mathcal{B}F_\alpha|\right)^{-1} \sup_{\mathcal{E}}|F(0;\varepsilon)| \ .$$

At this point the proof of Theorem 2.7.1 goes through estimate-by-estimate and the analyticity in the parameter ε follows, as mentioned above, from Weierstrass theorem.

2.8.2. The nearly-integrable case: Recursive equations.

We apply, here, the result of the previous section[2.37] to the nearly-integrable case

$$(2.8.4)\quad H(x,y;\varepsilon) = H_0(y)+\varepsilon H_1(x,y;\varepsilon)\ ,\ x\in\mathbb{T}^d_{\bar\xi}\ ,\quad y\in D^d_r(y_0)\ ,\quad \varepsilon\in D^1_{\varepsilon_0}(0)\ ,$$

under Arnold's iso-energetical non-degeneracy assumption on the energy level $E_0 := H_0(y_0)$, i.e.,

$$\det\mathcal{A}_0\neq 0\ ,\quad \mathcal{A}_0 := \begin{pmatrix} H_0''(y_0) & \omega_0 \\ \omega_0 & 0\end{pmatrix},$$

$$(2.8.5)\qquad \omega_0 := H_0'(y_0)\ ,\quad H_0(y_0) =: E_0\ .$$

Assume that the vector $\omega := \omega_0$ is (γ_0,τ)-Diophantine (in particular, we let $a(\varepsilon)$ in (2.8.1) to be zero[2.38]); assume that H is real-analytic on $\mathbb{T}^d_{\bar\xi}\times D^d_r(y_0)\times D^1_{\varepsilon_0}(0)$ and let

$$(2.8.6)\qquad E(\varepsilon) := E_0+\varepsilon E_1+\cdots\ ,$$

be a real-analytic function of $\varepsilon\in D^1_{\varepsilon_0}(0)$. Now, we take as approximate KAM torus simply $(u,v,\omega) := (0,y_0,\omega_0)$. It is then easy to check that, *fixed* $0<\xi_\infty<\bar\xi$, *if one takes* ε_0 *small enough*[2.39] *then* (2.7.8) *can be met and there exists a real-analytic solution* $(\tilde u,\tilde v) = (\tilde u(\theta;\varepsilon),\tilde v(\theta;\varepsilon))$ *of the system* (2.8.2), *which satisfies* (2.8.3) *and is the* ε*-analytic continuation of the unperturbed solution* $(0,y_0,\omega_0)$.

Thus, the functions $\tilde u$, $\tilde v$ and the number $\tilde a$ admit convergent power series expansions

$$(2.8.7)\qquad \tilde u = \sum_{j=1}^\infty \varepsilon^j u_j(\theta)\ ,\qquad \tilde v = y_0+\sum_{j=1}^\infty \varepsilon^j v_j(\theta)\ ,\qquad \tilde a = \sum_{j=1}^\infty \varepsilon^j a_j\ ,$$

in the complex domain $\mathcal{E} := \{\varepsilon\in\mathbb{C} : |\varepsilon|\le\varepsilon_0\}$; such series will be called *the iso-energetic Lindstedt series* for the Hamiltonian system associated to (2.8.4), (2.8.5).

[2.37] With: $m=1$ and $\mathcal{E}\equiv D^1_{\varepsilon_0}(0)$.

[2.38] The assumption that ω_0 is Diophantine, thanks to the iso-energetic non-degeneracy (2.8.5), can always be met by a slight change of y_0 on $H_0^{-1}(E_0)$. In fact, as discussed in point (ii) of the Remark 2.3.10, (2.8.5) implies that the map $y\in H_0^{-1}(E_0)\to\pi\circ H_0'(y)$ (π being the canonical projection of $\mathbb{R}^d$ onto $\mathbb{P}^{d-1}$) is a local diffeomorphism and this implies that the set of Diophantine points on $H_0^{-1}(E_0)$ is dense in any small neighborhood of y_0.

[2.39] Notice that $f = -\varepsilon\partial_y H_1(\cdot,y_0)$, $g = \varepsilon\partial_x H_1(\cdot,y_0)$, $h = \varepsilon(H_1(0,y_0)-E_1)+\sum_{j=2}^\infty E_j\varepsilon^j$ so that μ in (2.7.6) is proportional to ε_0.

The iso-energetic Lindstedt series satisfy a set of "linear recursive equations": denote by $[\cdot]_k$ the projection onto the kth coefficient of an ε-power series,

$$[\,\cdot\,]_k := \frac{1}{k!}\left.\frac{d^k(\cdot)}{d\varepsilon^k}\right|_{\varepsilon=0} ,$$

and, for $k \geq 1$, define (recall that $\omega = \omega_0$)

$$\begin{aligned}
X_k \;&:=\; -\sum_{j=1}^{k-1} a_j D_\omega u_{k-j} + \left[H_0'\Big(y_0 + \sum_{j=1}^{k-1}\varepsilon^j v_j\Big)\right]_k \\
&\qquad + \left[H_{1,y}\Big(\theta + \sum_{j=1}^{k-1}\varepsilon^j u_j, y_0 + \sum_{j=1}^{k-1}\varepsilon^j v_j\Big)\right]_{k-1} , \\
Y_k \;&:=\; -\sum_{j=1}^{k-1} a_j D_\omega v_{k-j} - \left[H_{1,x}\Big(\theta + \sum_{j=1}^{k-1}\varepsilon^j u_j, y_0 + \sum_{j=1}^{k-1}\varepsilon^j v_j\Big)\right]_{k-1} , \\
Z_k \;&:=\; -\left[H_0\Big(y_0 + \sum_{j=1}^{k-1}\varepsilon^j v_j(0)\Big)\right]_k - \left[H_1\Big(0, y_0 + \sum_{j=1}^{k-1}\varepsilon^j v_j(0)\Big)\right]_{k-1} + E_k .
\end{aligned}$$

(2.8.8)

REMARK 2.8.2. X_k and Y_k are real-analytic function of $\theta \in \mathbb{T}^d_{\xi_\infty}$, while Z_k is just a real number; X_k, Y_k and Z_k depend upon $\{a_j\}$, $\{u_j\}$, $\{v_j\}$ for $1 \leq j \leq k-1$ only. When $k = 1$ the sums over j, in the formulae (2.8.8), are absent.

PROPOSITION 2.8.3. *Let, for $k \geq 1$, X_k, Y_k and Z_k as in (2.8.8). Then $\langle Y_k \rangle = 0$ and, if one sets*

$$\begin{pmatrix} \bar{v}_k \\ -a_k \end{pmatrix} = \mathcal{A}_0^{-1} \begin{pmatrix} -\langle X_k \rangle \\ Z_k \end{pmatrix} , \tag{2.8.9}$$

then

$$v_k = D_\omega^{-1} Y_k + \bar{v}_k . \tag{2.8.10}$$

Furthermore

$$\Big\langle H_0''(y_0) v_k + X_k - a_k \omega \Big\rangle = 0 , \tag{2.8.11}$$

and, if we define

$$\breve{u}_k := D_\omega^{-1}\Big(H_0''(y_0) v_k + X_k - a_k \omega\Big) , \tag{2.8.12}$$

then

$$u_k = \breve{u}_k - \breve{u}_k(0) . \tag{2.8.13}$$

Notice that the identity $\langle Y_k \rangle = 0$ together with (2.8.11) may be interpreted as *compatibility conditions*, while (2.8.9), (2.8.10), (2.8.12) and (2.8.13) may be viewed as *definitions* of a_k, u_k and v_k.

The proof of this Proposition 2.8.3 is, at this point, trivial: plug the ε-expansion of $\tilde{a}$, $\tilde{u}$ and $\tilde{v}$ into the system (2.8.2) and expand in ε; the compatibility conditions are immediately checked by taking averages over $\mathbb{T}^d$ of the obtained relations. □

CHAPTER 3

The Restricted, Circular, Planar Three-body Problem

3.1. The restricted three-body problem

Roughly speaking the *restricted three-body problem* is the problem of describing the bounded motions of a "zero-mass" body subject to the gravitational field generated by an assigned two-body system[3.1]. To describe mathematically such system, let P_0, P_1, P_2 be three bodies ("point masses") with masses m_0, m_1, m_2 interacting only through the gravitational attraction. If $u^{(i)} \in \mathbb{R}^3$, $i = 1, 2, 3$, denote the position of the bodies in some (inertial) reference frame (and assuming, without loss of generality, that the gravitational constant is one[3.2]), the Newton equations for this system have the form

$$\begin{aligned} \frac{d^2u^{(0)}}{dt^2} &= -\frac{m_1(u^{(0)} - u^{(1)})}{|u^{(1)} - u^{(0)}|^3} - \frac{m_2(u^{(0)} - u^{(2)})}{|u^{(2)} - u^{(0)}|^3}, \\ \frac{d^2u^{(1)}}{dt^2} &= -\frac{m_0(u^{(1)} - u^{(0)})}{|u^{(1)} - u^{(0)}|^3} - \frac{m_2(u^{(1)} - u^{(2)})}{|u^{(2)} - u^{(1)}|^3}, \\ \frac{d^2u^{(2)}}{dt^2} &= -\frac{m_0(u^{(2)} - u^{(0)})}{|u^{(2)} - u^{(0)}|^3} - \frac{m_1(u^{(2)} - u^{(1)})}{|u^{(2)} - u^{(1)}|^3}. \end{aligned} \tag{3.1.1}$$

The *restricted three-body problem (with "primary bodies" P_0 and P_1) is*, by definition, *the problem of studying the bounded motions of the system* (3.1.1) *after having set* $m_2 = 0$, i.e., of the system

$$\begin{aligned} \frac{d^2u^{(0)}}{dt^2} &= -\frac{m_1(u^{(0)} - u^{(1)})}{|u^{(1)} - u^{(0)}|^3}, \qquad \frac{d^2u^{(1)}}{dt^2} = -\frac{m_0(u^{(1)} - u^{(0)})}{|u^{(1)} - u^{(0)}|^3} \\ \frac{d^2u^{(2)}}{dt^2} &= -\frac{m_0(u^{(2)} - u^{(0)})}{|u^{(2)} - u^{(0)}|^3} - \frac{m_1(u^{(2)} - u^{(1)})}{|u^{(1)} - u^{(2)}|^3}. \end{aligned} \tag{3.1.2}$$

Notice that the equations for the two primaries P_0 and P_1 *decouple* and describe an *unperturbed two-body system*, which can be solved and the solution can be plugged into the equation for $u^{(2)}$, which becomes a second-order, *periodically forced* equation in $\mathbb{R}^3$.

3.2. Delaunay action-angle variables for the two-body problem

In this section we review the construction of the classical Delaunay [**46**] action-angle variables for the two-body problem.

[3.1] For general references, see, e.g., [**129**], [**120**].

[3.2] This amounts to rescale the time.

The equations of motion of two bodies P_0 and P_1 of masses m_0 and m_1, interacting through gravitation (with gravitational constant equal to one) are given (as in the first line of (3.1.2)) by

$$\begin{aligned} \frac{d^2u^{(0)}}{dt^2} &= -\frac{m_1(u^{(0)}-u^{(1)})}{|u^{(1)}-u^{(0)}|^3} \ , \\ \frac{d^2u^{(1)}}{dt^2} &= -\frac{m_0(u^{(1)}-u^{(0)})}{|u^{(1)}-u^{(0)}|^3} \ , \qquad u^{(i)} \in \mathbb{R}^3 \ . \end{aligned} \tag{3.2.1}$$

As everybody knows, the total energy, momentum and angular momentum are preserved. We shall therefore fix an *inertial frame* $\{k_1, k_2, k_3\}$, with origin in the center of mass and with k_3-axis parallel to the total angular momentum. In such frame we have

$$u_3^{(0)} \equiv 0 \equiv u_3^{(1)} \ , \qquad m_0 u^{(0)} + m_1 u^{(1)} = 0 \ . \tag{3.2.2}$$

We pass to a *heliocentric frame* by letting,

$$(x, 0) := u^{(1)} - u^{(0)} \ , \qquad x \in \mathbb{R}^2 \ . \tag{3.2.3}$$

In view of (3.2.1) and (3.2.2), the equations for x become

$$\ddot{x} = -M\frac{x}{|x|^3} \ , \qquad M := m_0 + m_1 \ . \tag{3.2.4}$$

This equation is Hamiltonian: let $\mu > 0$ and set

$$H_{\mathrm{Kep}}(x, X) := \frac{|X|^2}{2\mu} - \frac{\mu M}{|x|} \ , \qquad X := \mu \dot{x} \ , \tag{3.2.5}$$

then (3.2.4) is equivalent to the Hamiltonian equation associated to H_{Kep} with respect to the standard symplectic form $dx \wedge dX$, the phase space being $\mathbb{R}^2 \backslash \{0\} \times \mathbb{R}^2$; the (free) parameter μ is traditionally chosen as the "reduced mass" $m_0 m_1 / M$.

The motion in the u-coordinates is recovered (via (3.2.2) and (3.2.3)) by the relation

$$u^{(0)} = \Big(\frac{-m_1}{M}x, 0\Big) \ , \qquad u^{(1)} = \Big(\frac{m_0}{M}x, 0\Big) \ ,$$

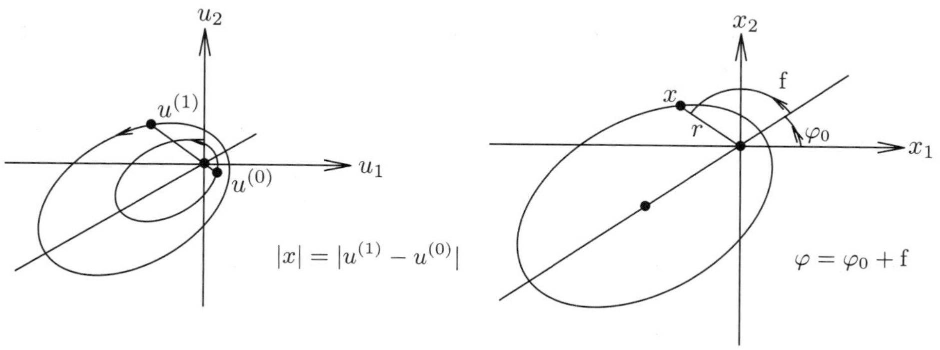

FIGURE 2. The geometry of the Kepler two–body problem.

The dependence of $H_{\rm Kep}$ on x through the absolute value suggests to introduce polar coordinates in the x-plane: $x = r(\cos\varphi, \sin\varphi)$ and, in order to get a symplectic transformation, one is led to the symplectic map $\phi_{\rm pc} : ((r,\varphi),(R,\Phi)) \to (x, X)$ given by

$$
(3.2.6) \qquad \phi_{\rm pc} : \quad \begin{cases} x \;= r(\cos\varphi, \sin\varphi)\ , \\ X = \Big(R\cos\varphi - \frac{\Phi}{r}\sin\varphi, R\sin\varphi + \frac{\Phi}{r}\cos\varphi\Big)\ , \\ dx_1 \wedge dX_1 + dx_2 \wedge dX_2 = dr \wedge dR + d\varphi \wedge d\Phi\ . \end{cases}
$$

The variables r and φ are commonly called, in Celestial Mechanics, *the orbital radius* and the *longitude of the planet* P_1.

In the new symplectic variables the Hamiltonian $H_{\rm Kep}$ takes the form

$$
H_{\rm pc}(r,\varphi,R,\Phi) := H_{\rm Kep} \circ \phi_{\rm pc}(r,\varphi,R,\Phi) = \frac{1}{2\mu}\Big(R^2 + \frac{\Phi^2}{r^2}\Big) - \frac{\mu M}{r}\ .
$$

The variable φ is *cyclic* (i.e., $\partial H_{\rm pc}/\partial\varphi = 0$ so that $\Phi \equiv$ const), showing that the system with Hamiltonian $H_{\rm pc}$ is actually a *one-degree-of-freedom Hamiltonian system* (in the symplectic variables (r, R)), and is therefore *integrable*. The momentum variable Φ conjugated to φ is an integral of motion and

$$
\dot\varphi = \frac{\partial H_{\rm pc}}{\partial\Phi} = \frac{\Phi}{\mu r^2} \qquad \Longrightarrow \qquad \Phi = \mu r^2 \dot\varphi \equiv \text{const}\ .
$$

REMARK 3.2.1. The total angular momentum, C, in the inertial frame (and referred to the center of mass) is given by[3.3]

$$
C = m_0 u^{(0)} \times \dot u^{(0)} + m_1 u^{(1)} \times \dot u^{(1)}\ .
$$

Taking into account the inertial relation $m_0 u^{(0)} = -m_1 u^{(1)}$ one finds that

$$
C = \frac{m_0 m_1}{M}(x,0) \times (\dot x, 0) = \frac{m_0 m_1}{M\mu}\,(x,0) \times (X,0)\ ,
$$

and the evaluation of the angular momentum in polar coordinates shows that

$$
C = \pm\, k_3\, \frac{m_0 m_1}{M} r^2 \dot\varphi = \pm\, k_3 \frac{m_0 m_1}{M\mu}\Phi\ ;
$$

[3.3] "$\times$" denotes, here, the standard "vector product" in $\mathbb{R}^3$.

thus if μ is chosen to be the reduced mass $\frac{m_0 m_1}{M}$, then Φ is exactly the absolute value of the total angular momentum.

The analysis of the (r, R) motion is standard: introducing the "effective potential"

$$V_{\text{eff}}(r) := V_{\text{eff}}(r;\Phi) := \frac{\Phi^2}{2\mu r^2} - \frac{\mu M}{r} ,$$

one is led to the "effective Hamiltonian" (parameterized by Φ)

$$H_{\text{eff}} = \frac{R^2}{2\mu} + V_{\text{eff}}(r) \qquad (R = \mu\dot{r}) .$$

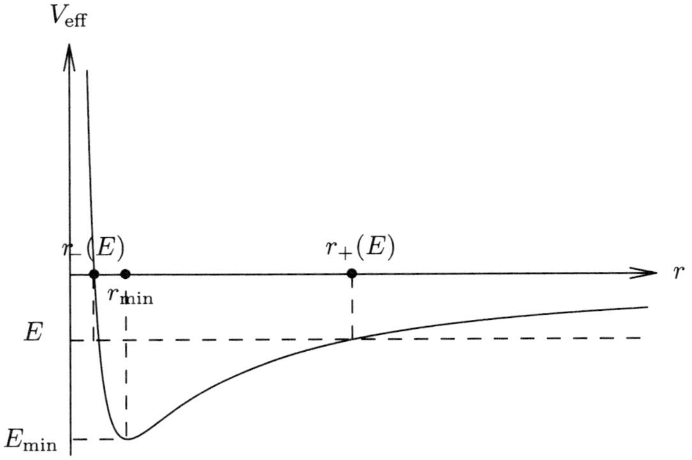

FIGURE 3. The effective potential of the two–body problem.

The motion on the energy level $H_{\text{eff}}^{-1}(E)$ is bounded (and periodic) if and only if

$$E \in [E_{\min}, 0) , \qquad E_{\min} := V_{\text{eff}}(r_{\min}) = -\frac{\mu^3 M^2}{2\Phi^2} , \qquad r_{\min} := \frac{\Phi^2}{\mu^2 M} . \tag{3.2.7}$$

For $E \in (E_{\min}, 0)$ the period $T(E)$ is given by

$$T(E) = 2\int_{r_-(E)}^{r_+(E)} \frac{dr}{\sqrt{\frac{2}{\mu}(E - V_{\text{eff}}(r))}} , \tag{3.2.8}$$

where $r_\pm(E) = r_\pm(E;\Phi)$ are the two positive roots of $E - V_{\text{eff}}(r) = 0$, i.e.,

$$E - V_{\text{eff}}(r) =: \frac{-E}{r^2}(r_+ - r)(r - r_-) ,$$

$$r_\pm(E;\Phi) = \frac{\mu M \pm \sqrt{(\mu M)^2 + \frac{2E\Phi^2}{\mu}}}{-2E} . \tag{3.2.9}$$

The integral in (3.2.8) is readily computed yielding Kepler's second law

$$T(E) = 2\pi\ M\left(\frac{\mu}{-2E}\right)^{3/2} .$$

Let us now integrate the motion in the (r, φ) coordinates. The equations of motion in such coordinates are given by

$$\dot\varphi = \frac{\Phi}{\mu r^2}\ , \qquad \dot r^2 = \frac{2}{\mu}(E - V_{\rm eff}(r)) \ . \tag{3.2.10}$$

By symmetry arguments, it is enough to consider the motion for $0 \le t \le T(E)/2$; furthermore, we shall choose the initial time so that $r(0) = r_-$ (i.e., at the initial time the system is at the "*perihelion*"): the corresponding angle will be a certain φ_0 and we shall make the (trivial) change of variables

$$\varphi = \varphi_0 + {\rm f}\ , \qquad \text{so that} \quad r(0) = r_-(E)\ , \quad {\rm f}(0) = 0\ . \tag{3.2.11}$$

The angle f is commonly called the *true anomaly*; the angle φ_0 (i.e., the angle between the perihelion line, joining the foci of the ellipse and the x_1 axis) is called *the argument of the perihelion* (compare with figure 2).

Equations (3.2.10) become

$$\begin{cases} \dot{\rm f} = \frac{\Phi}{\mu r^2}\ , & {\rm f}(0) = 0 \\ \dot r^2 = \frac{2}{\mu}(E - V_{\rm eff}(r))\ , & r(0) = r_-(E)\ . \end{cases}$$

Eliminating time (for $t \in (0, T(E))$, $\dot r > 0$) we find (recall the definitions in (3.2.7))

$${\rm f} = \Phi \int_{r_-(E)}^{r} \frac{d\rho/\rho^2}{\sqrt{2\mu(E - V_{\rm eff}(\rho))}} = {\rm Arccos}\ \frac{\frac{r_{\rm min}}{r} - 1}{\sqrt{1 - \frac{E}{E_{\rm min}}}}\ . \tag{3.2.12}$$

Setting

$$e := \sqrt{1 - \frac{E}{E_{\rm min}}}\ , \qquad p := r_{\rm min}\ , \tag{3.2.13}$$

we get the classical focal equation

$$r = \frac{p}{1 + e\cos{\rm f}} = \frac{p}{1 + e\cos(\varphi - \varphi_0)}\ , \tag{3.2.14}$$

which shows that P_0 and P_1 describe two ellipses of eccentricity $e \in (0, 1)$ with common focus in the center of mass (first Kepler law).

If $a \ge b > 0$ denote the semi-axis of the ellipse, from (3.2.14) it follows immediately that

$$r_\pm = \frac{p}{1 \mp e}\ , \quad r_+ + r_- = 2a\ , \quad p = a(1 - e^2)\ , \quad r_\pm = a(1 \pm e)\ . \tag{3.2.15}$$

From the geometry of the ellipse (see Appendix A and in particular (A.0.2) and figure 7) one knows that

$$r = a(1 - e\cos u)\ , \tag{3.2.16}$$

where u is the so-called *eccentric anomaly*. Then, from the definition of $E_{\rm min}$, (3.2.7), the expression for $E - V_{\rm eff}$ in (3.2.9), the relations (3.2.13) and (3.2.15), one finds

$$E_{\rm min} = -\frac{\mu M}{2p}\ , \quad E = -\frac{\mu M}{2a}\ , \quad E - V_{\rm eff} = \frac{\mu M}{2a}\left(\frac{e\sin u}{1 - e\cos u}\right)^2\ . \tag{3.2.17}$$

REMARK 3.2.2. The *circular motion* for the two-body problem is obtained for the minimal value of the energy $E = E_{\min} = -\frac{\mu^3 M^2}{2\Phi^2}$. In such a case

$$e = 0 \ , \qquad r \equiv p = r_{\min} = \frac{\Phi^2}{\mu^2 M} \ ; \tag{3.2.18}$$

the constant angular velocity and the period are respectively given by

$$\omega_{\text{circ}} = \frac{\mu^3 M^2}{\Phi^3} \ , \qquad T_{\text{circ}} = 2\pi \frac{\Phi^3}{\mu^3 M^2} \ . \tag{3.2.19}$$

Eliminating Φ in (3.2.18) and (3.2.19) one gets

$$\omega_{\text{circ}} = \sqrt{\frac{M}{r^3}} \ , \qquad T_{\text{circ}} = 2\pi \sqrt{\frac{r^3}{M}} \ .$$

The motion in the x-variables is given by

$$x(t) = r\Big(\cos(\varphi_0 + \omega_{\text{circ}}\ t), \sin(\varphi_0 + \omega_{\text{circ}}\ t)\Big) \ . \tag{3.2.20}$$

We turn to the construction of the *action-angle variables*. For $E \in (E_{\min}, 0)$, denote by S_E the curve (energy level) $\{(r, R) : H_{\text{eff}}(r, R) = E\}$ (at a fixed value of Φ). The area $A(E)$ encircled by such a curve in the (r, R)-plane is given by

$$A(E) = 2\int_{r_-(E)}^{r_+(E)} \sqrt{2\mu\Big(E - V_{\text{eff}}(r)\Big)}\, dr = 2\pi\ \mu M \sqrt{\frac{\mu}{-2E}} - 2\pi\Phi \ .$$

Thus, (by the theorem of Liouville-Arnold) the action variable is given by

$$I(E) = \frac{A(E)}{2\pi} = \mu M \sqrt{\frac{\mu}{-2E}} - \Phi \ ,$$

which, inverted, gives the form of the Hamiltonian H_{eff} in the action-angle variables (θ, I) (and parameterized by Φ):

$$h(I) := h(I; \Phi) := -\frac{\mu^3 M^2}{2(I + \Phi)^2} \ .$$

Furthermore (again by Liouville-Arnold), the symplectic transformation between (r, R) (in a neighborhood of a point with $R > 0$) and the action variables, (θ, I), for the Hamiltonian H_{eff} is *generated* by the generating function[3.4]

$$S_0(I, r; \Phi) := \int_{(r_-(h(I)),0)}^{(r, R_+(r;I))} R dr \ , \qquad R_+(r; I) := \sqrt{2\mu\big(h(I) - V_{\text{eff}}(r)\big)} \ ,$$

where the integration is performed over the curve $S_{h(I)}$ oriented *clockwise*: the orientation of $S_{h(I)}$ and the choice of the base point as $\big(r_-(h(I)), 0\big)$ is done so that an integration over the closed curve gives $+A(E)$ and so that $\theta = 0$ corresponds to the perihelion position.

[3.4] Recall that the dependence upon Φ is hidden in r_- and V_{eff}.

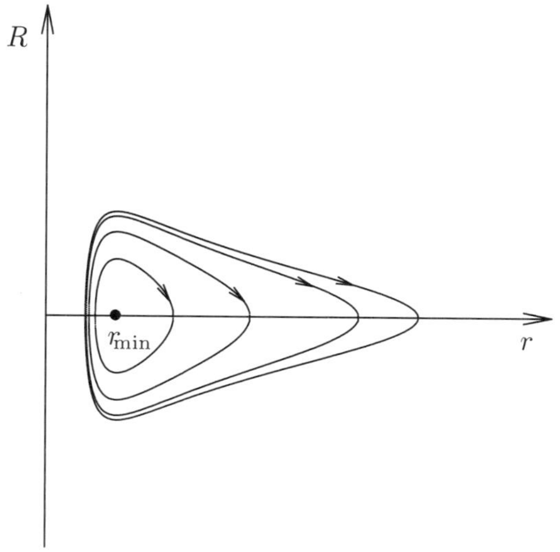

FIGURE 4. Level curves of the effective Hamiltonian for $E_{\min} < E < 0$.

The full symplectic transformation (in the four dimensional phase space of $H_{\rm pc}$)

$$\phi_{\rm aa}: \quad \begin{cases} (\theta, \psi, I, J) \to (r, \varphi, R, \Phi) \\ d\theta \wedge dI + d\psi \wedge dJ = dr \wedge dR + d\varphi \wedge d\Phi \end{cases}$$

will then be generated by the generating function

$$S_1(I, J, r, \varphi) := S_0(I, r; J) + J\varphi \ , \qquad (J = \Phi) \ .$$

The form of $h(I)$ suggests to introduce one more (linear, symplectic) change of variables given by

$$\phi_{\rm lin}^{-1}: \quad \begin{cases} \Lambda = I + J \ , & \Gamma = J \ , \\ \lambda = \theta \ , & \gamma = \psi - \theta \ . \end{cases}$$

The variables $(\lambda, \gamma, \Lambda, \Gamma)$ are the celebrated *Delaunay variables for the two-body problem.* If we set

$$\phi_{\rm D} := \phi_{\rm pc} \circ \phi_{\rm aa} \circ \phi_{\rm lin} \tag{3.2.21}$$

by the above analysis we get

$$h_{\rm Kep} \circ \phi_{\rm D}(\lambda, \gamma, \Lambda, \Gamma) = h_{\rm Kep}(\Lambda) := -\frac{\mu^3 M^2}{2\Lambda^2} \ . \tag{3.2.22}$$

The symplectic transformation $\phi_{\rm aa} \circ \phi_{\rm lin}$ is generated by $(\Gamma = J = \Phi)$

$$\begin{aligned} S_2(\Lambda, \Gamma, r, \varphi) &:= S_0(\Lambda - \Gamma, r; \Gamma) + \Gamma\varphi \\ &= \int_{r_-(h_{\rm Kep}(\Lambda))}^{r} \sqrt{-\frac{\mu^4 M^2}{\Lambda^2} + \frac{2\mu^2 M}{\rho} - \frac{\Gamma^2}{\rho^2}} \, d\rho + \Gamma\varphi \\ &= \sqrt{2\mu} \int_{r_-(h_{\rm Kep}(\Lambda))}^{r} \sqrt{h_{\rm Kep}(\Lambda) - V_{\rm eff}(\rho; \Gamma)} \, d\rho \ + \Gamma\varphi \ . \end{aligned}$$

Replacing E by $h_{\rm Kep}(\Lambda)$ and Φ with Γ in the expression for the eccentricity e in (3.2.13) (recall the definition of $E_{\rm min}$ in (3.2.7)) one finds

$$e = e(\Lambda, \Gamma) = \sqrt{1 - \Big(\frac{\Gamma}{\Lambda}\Big)^2} \; . \tag{3.2.23}$$

Recalling also the second relation in (3.2.17), one finds

$$a = \frac{\Lambda^2}{\mu^2 M} \; , \qquad \Lambda = \mu\sqrt{Ma} \; . \tag{3.2.24}$$

REMARK 3.2.3. Recall that, by (3.2.7) (and fixing suitably the direction of the k_3-axis),

$$\Gamma = \Phi = \frac{\mu M}{m_0 m_1}|C| \; , \qquad C := \text{ total angular momentum } ,$$

so that

$$\Gamma > 0 \; .$$

Recall also that

$$E_{\rm min} = -\frac{\mu^3 M^2}{2\Gamma^2} \; ,$$

so that $E > E_{\rm min}$ means (by (3.2.22))

$$\Gamma < \Lambda \; .$$

The momentum space is therefore the *positive cone* $\{0 < \Gamma < \Lambda\}$.

The angle λ is computed from the generating function S_2:

$$\begin{aligned}
\lambda &= \frac{\partial S_2}{\partial \Lambda} = \sqrt{\frac{\mu}{2}}\,\frac{\mu^3 M^2}{\Lambda^3}\int_{r_-}^{r} \frac{d\rho}{\sqrt{h_{\rm Kep}(\Lambda) - V_{\rm eff}(\rho;\Gamma)}} \\
&\overset{(3.2.24)}{=} \sqrt{\frac{\mu M}{2a}}\,\frac{1}{a}\int_{r_-}^{r} \frac{d\rho}{\sqrt{h_{\rm Kep}(\Lambda) - V_{\rm eff}(\rho;\Gamma)}} \\
&\overset{(3.2.17)}{=} \frac{1}{a}\int_{r_-}^{r} \frac{1 - e\cos u}{e\sin u}\, d\rho \\
&\overset{(3.2.16)}{=} \int_0^u (1 - e\cos u)du \\
&= u - e\sin u \\
&= 2\pi\frac{\text{Area}(\mathcal{E}(\mathrm{f}))}{\text{Area}(\mathcal{E}(2\pi))} \; ,
\end{aligned} \tag{3.2.25}$$

where (compare Appendix A), $\mathcal{E}(\mathrm{f})$ is the area (on the ellipse (3.2.14)) "spanned by the orbital radius":

$$\mathcal{E}(\mathrm{f}) := \{x = x(r', \mathrm{f}') : 0 \le r' \le r(\mathrm{f}) \; , 0 \le \mathrm{f}' \le \mathrm{f}\} \; ;$$

we have also used the fact that ρ as a function of $u \in [0, \pi]$ is a strictly increasing function and that $\rho(0) = r_-$.

In view of (3.2.25), λ is called *the mean anomaly*. Analogously, *the angle* γ *is recognized to be the argument of the perihelion* φ_0 introduced above (just before

Remark 3.2.2):

$$\begin{aligned}\gamma &= \frac{\partial S_2}{\partial \Gamma} = \varphi - \Gamma \int_{r_-}^{r} \frac{1}{\sqrt{2\mu(h_{\rm Kep}(\Lambda) - V_{\rm eff}(\rho))}} \frac{d\rho}{\rho^2} \\ &\overset{(3.2.12)}{=} \varphi - \mathrm{f} \\ &\overset{(3.2.11)}{=} \varphi_0 \ . \end{aligned} \tag{3.2.26}$$

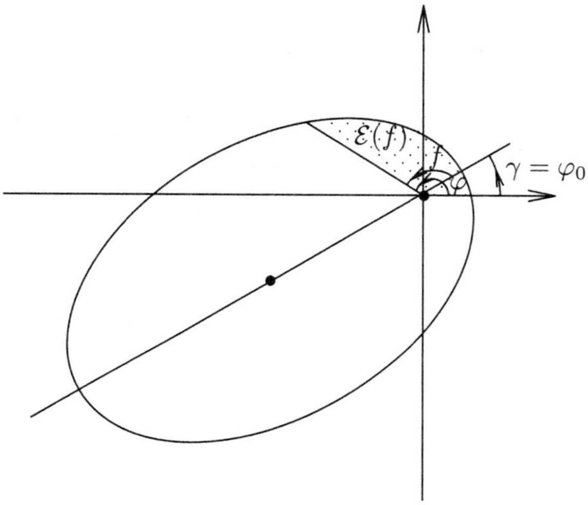

FIGURE 5. The Delaunay angles.

We conclude this classical section by giving *analytical expressions for the eccentric anomaly* u, *the true anomaly* f, *the longitude* φ *and the orbital radius* r *in terms of the Delaunay variables.*

The (Kepler) equation

$$\lambda = u - e \sin u \ ,$$

(see (3.2.25)) can be inverted, for $e \in [0, 1)$, as

$$\begin{aligned} u &= u_0(\lambda, e) := \lambda + e\tilde{u}(\lambda, e) \\ &= \lambda + e \sin\lambda + \frac{e^2}{2} \sin 2\lambda + \frac{e^3}{8}\Big(-\sin\lambda + 3\sin 3\lambda\Big) + \cdots \ , \end{aligned} \tag{3.2.27}$$

where $\tilde{u}$ is real-analytic in $\lambda \in \mathbb{T}$ and $e \in [0, 1)$; via (3.2.23), $e = e(\Lambda, \Gamma) = \sqrt{1 - (\Gamma/\Lambda)^2}$, the relation (3.2.27) yields an analytic expression of the eccentric anomaly as a function of the Delaunay variables λ, Λ, Γ.

From the geometry of the ellipse it follows that (compare (??) in Appendix A)

$$\tan\frac{\mathrm{f}}{2} = \sqrt{\frac{1+e}{1-e}} \tan\frac{u}{2} \ ,$$

which can be written, for $e \in [0,1)$, as

$$\begin{aligned} \mathrm{f} &= \hat{\mathrm{f}}_0(u,e) := u + e\hat{\mathrm{f}}(u,e) \\ &= u + e\,\sin u + \frac{e^2}{4}\,\sin 2\,u + \frac{e^3}{12}\Big(3\,\sin u + \sin 3\,u\Big) + \cdots\,, \end{aligned} \tag{3.2.28}$$

where $\hat{\mathrm{f}}$ is analytic in $u \in \mathbb{T}$ and $e \in [0,1)$. Through (3.2.27) and (3.2.23), the expression (3.2.28) yields an analytic expression of the true anomaly f in terms of λ, Λ, Γ:

$$\begin{aligned} \mathrm{f} &= \mathrm{f}_0(\lambda,e) := \hat{\mathrm{f}}_0(u_0(\lambda,e),e) =: \lambda + e\tilde{\mathrm{f}}(\lambda,e) \\ &= \lambda + 2\,e\,\sin\lambda + \frac{5}{4}\,e^2\,\sin 2\,\lambda + \frac{e^3}{12}\Big(-3\,\sin\lambda + 13\,\sin 3\,\lambda\Big) + \cdots\,. \end{aligned} \tag{3.2.29}$$

As above $e = e(\Lambda, \Gamma)$.

The longitude φ by (3.2.26) is simply $\varphi = \gamma + \mathrm{f}$ and can, therefore, be expressed as a function of $\lambda, \gamma, \Lambda, \Gamma$.

From the geometry of the ellipse it follows that (compare (A.0.2) in Appendix A) r is related to a, e and u by

$$r = a(1 - e\cos u)\,.$$

Thus by (3.2.24), (3.2.23) and (3.2.27) we find

$$\begin{aligned} \frac{r}{a} &= \frac{r_0(\lambda,e)}{a} \\ &:= 1 - e\cos u_0(\lambda,e) \\ &= 1 - e\,\cos\lambda + \frac{e^2}{2}\Big(1 - \cos 2\,\lambda\Big) + \frac{3}{8}\,e^3\Big(\cos\lambda - \cos 3\,\lambda\Big) + \cdots\,, \end{aligned} \tag{3.2.30}$$

where $e = e(\Lambda,\Gamma)$ and $a = a(\Lambda) := \Lambda^2/(\mu^2 M)$ (see (3.2.23) and (3.2.24)).

3.3. The restricted, circular, planar three-body problem viewed as nearly-integrable Hamiltonian system

Let us go back to (3.1.2). Since we shall study the *planar three-body problem*, we *assume that the motion takes place on the plane hosting the Keplerian motion of P_0 and P_1*. This amounts to require

$$u_3^{(i)} \equiv 0 \equiv \dot{u}_3^{(i)}\,, \qquad i = 0,1,2\,. \tag{3.3.1}$$

Observe that, since we are considering the *restricted* problem (i.e. we have set in (3.1.1) $m_2 = 0$), the "conservation laws" are those of the two-body system $P_0 - P_1$: in particular the total angular momentum is parallel to the u_3-axis (consistently with (3.3.1)) and the center of mass (and hence the origin of the u-frame) is simply

$$m_0 u^{(0)} + m_1 u^{(1)} = 0\,. \tag{3.3.2}$$

Next, we pass, as in § 3.2, to *heliocentric coordinates*:

$$(x^{(1)},0) := u^{(1)} - u^{(0)}\,, \qquad (x^{(2)},0) := u^{(2)} - u^{(0)}\,, \qquad \Big(x^{(1)}, x^{(2)} \in \mathbb{R}^2\Big)\,,$$

which transform (3.1.2) into

$$\ddot{x}^{(1)} := -M_0 \frac{x^{(1)}}{|x^{(1)}|^3} , \qquad M_0 := m_0 + m_1 , \tag{3.3.3}$$

$$\ddot{x}^{(2)} := -m_0 \frac{x^{(2)}}{|x^{(2)}|^3} - m_1 \frac{x^{(1)}}{|x^{(1)}|^3} - m_1 \frac{x^{(2)} - x^{(1)}}{|x^{(2)} - x^{(1)}|^3} . \tag{3.3.4}$$

In view of (3.3.2) the motion in the original u-coordinates is related to the motion in the heliocentric coordinates by

$$u^{(0)} = \Big(\frac{-m_1}{M_0} x^{(1)}, 0 \Big) , \quad u^{(1)} = \Big(\frac{m_0}{M_0} x^{(1)}, 0 \Big) , \quad u^{(2)} = \Big(x^{(2)} - \frac{m_1}{M_0} x^{(1)}, 0 \Big) .$$

The equation in (3.3.3) describes the decoupled two-body system $P_0 - P_1$, which has been discussed in § 3.2.

In the restricted, circular, planar three-body problem such motion is assumed to be circular.

It is convenient to *fix the measure units for lengths and masses so that the (fixed) distance between the two primary bodies is one and the sum of their masses is one*:

$$\text{dist}(P_0, P_1) = 1 , \qquad M_0 := m_0 + m_1 = 1 . \tag{3.3.5}$$

Recalling Remark 3.2.2, one sees that *the period of revolution of P_0 and P_1 around their center of mass (the "year") is, in such units, 2π*; the $x^{(1)}$-motion is simply (compare (3.2.20))

$$\hat{x}^{(1)}_{\text{circ}}(t) = x^{(1)}_{\text{circ}}(t_0 + t) := \Big(\cos(t_0 + t), \sin(t_0 + \ t) \Big) .$$

Even though the system of equations (3.3.3) *and* (3.3.4) is *not* a Hamiltonian system of equation, (3.3.3) and (3.3.4) taken *separately* are Hamiltonian: we have already seen that (3.3.3) represent just the equations of a two-body system; equations (3.3.4) represent a $2\frac{1}{2}$-degree-of-freedom Hamiltonian system with Hamiltonian

$$\begin{aligned} \tilde{H}_1(x^{(2)}, X^{(2)}, t) \ := \ & \frac{|X^{(2)}|^2}{2\mu} - \mu m_0 \frac{1}{|x^{(2)}|} + \mu m_1 \Big(x^{(2)} \cdot \hat{x}^{(1)}_{\text{circ}}(t) \Big) \\ & - \mu m_1 \frac{1}{|x^{(2)} - \hat{x}^{(1)}_{\text{circ}}(t)|} , \\ & (x^{(2)}, X^{(2)}) \in \mathbb{R}^2 \backslash \{0\} \times \mathbb{R}^2 , \ t \in \mathbb{T} , \end{aligned} \tag{3.3.6}$$

with respect to the standard symplectic form $dx^{(2)} \wedge dX^{(2)}$; here, $\mu > 0$ is a free parameter. To make the system (3.3.6) autonomous, we introduce a linear symplectic variable T conjugated to time $\tau = t$:

$$\begin{aligned} & \tilde{H}_1(x^{(2)}, X^{(2)}, \tau, T) \\ & := \frac{|X^{(2)}|^2}{2\mu} - \mu m_0 \frac{1}{|x^{(2)}|} + T + \mu m_1 \Big(x^{(2)} \cdot x^{(1)}_{\text{circ}}(\tau) \Big) - \mu m_1 \frac{1}{|x^{(2)} - x^{(1)}_{\text{circ}}(\tau)|} , \\ & \qquad (x^{(2)}, X^{(2)}) \in \mathbb{R}^2 \backslash \{0\} \times \mathbb{R}^2 , \ (\tau, T) \in \mathbb{T} \times \mathbb{R}. \end{aligned} \tag{3.3.7}$$

REMARK 3.3.1. In the limiting case of a primary body with mass $m_1 = 0$, the Hamiltonian $\tilde{H}_1$ describes a two-body system as in (3.2.5) with "total mass"

$$M = m_0 ,$$

reflecting the fact that the asteroid mass has been set equal to zero.

If the mass m_1 does not vanish but it is small compared to the mass of m_0, the system (3.3.7) may be viewed as a *nearly-integrable system.* This is more transparent if we use, for the integrable part, the Delaunay variables introduced in § 3.2 (see in particular (3.2.21)). Recall that the symplectic transformation $\phi_{\rm D}$, mapping the Delaunay variables to the original Cartesian variables, depends parametrically also on μ and M and that M is now m_0. Next, we choose the free parameter μ so as to make the Keplerian part equal to $-1/(2\Lambda^2)$ (see (3.3.9) below) and we introduce also a perturbation parameter ε closely related to the mass m_1 of the primary body:

$$\mu := \frac{1}{m_0^{2/3}} , \qquad \varepsilon := \frac{m_1}{m_0^{2/3}} = \frac{m_1}{(1-m_1)^{2/3}} . \tag{3.3.8}$$

Now, letting

$$\begin{aligned} &(\lambda,\gamma,\Lambda,\Gamma) = \phi_{\rm D}^{-1}(x^{(2)},X^{(2)}) , \\ &\hat\phi_{\rm D}\Big((\lambda,\gamma,\Lambda,\Gamma),(\tau,T)\Big) := \Big(\phi_{\rm D}(\lambda,\gamma,\Lambda,\Gamma),(\tau,T)\Big) , \end{aligned}$$

we find that[3.5]

$$\tilde H_2 := \tilde H_1 \circ \hat\phi_{\rm D} = -\frac{1}{2\Lambda^2} + T + \varepsilon\left(x^{(2)}\cdot x^{(1)}_{\rm circ}(\tau) - \frac{1}{|x^{(2)} - x^{(1)}_{\rm circ}(\tau)|}\right) , \tag{3.3.9}$$

where, of course, $x^{(2)}$ is now a function of the new symplectic variables.

Let us now analyze more in detail the perturbing function in (3.3.9).

Recalling the definition of φ in (3.2.6), one sees that the angle between the rays $(0,x^{(2)})$ and $(0,x^{(1)}_{\rm circ})$ is $\varphi-\tau$.

[3.5] Recall (3.2.22).

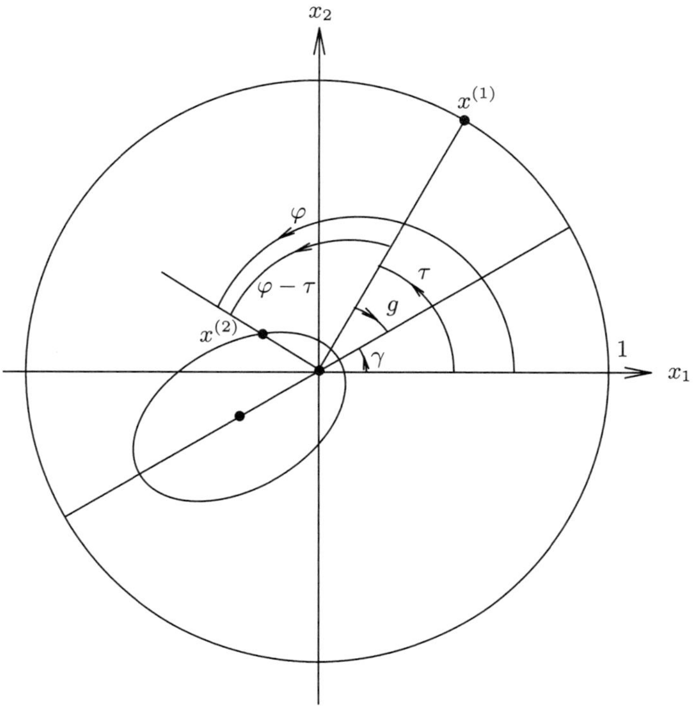

FIGURE 6. Angle variables for the RCP3BP.

Therefore, if we let

$$r_2 := |x^{(2)}| ,$$

we get

$$\tilde{H}_2 = -\frac{1}{2\Lambda^2} + T + \varepsilon \left(r_2 \cos(\varphi - \tau) - \frac{1}{\sqrt{1 + r_2^2 - 2r_2 \cos(\varphi - \tau)}} \right) .$$

Recall ((3.2.26)) that $\varphi = \gamma + \mathrm{f}$ and ((3.2.29)) that $\mathrm{f} = \mathrm{f}_0(\lambda, e) := \lambda + e\tilde{\mathrm{f}}(\lambda, e)$. Thus

$$\varphi - \tau = \mathrm{f} + \gamma - \tau = \lambda + \gamma - \tau + e\tilde{\mathrm{f}}(\lambda, e) .$$

Such relation suggests to make a new linear symplectic change of variables, by setting

$$\hat{\phi}_{\text{lin}}^{-1} : \quad \begin{cases} L = \Lambda , & G = \Gamma , & \hat{T} = T + \Gamma \\ \ell = \lambda , & g = \gamma - \tau , & \hat{\tau} = \tau . \end{cases}$$

Now, recalling (3.2.24), (3.3.8) and (3.3.5) we see that

$$a = L^2/(\mu^2 M) = m_0^{1/3} L^2 ,$$

so that, by (3.2.30), in the new symplectic variables, one finds

$$\varphi - \tau = \mathrm{f} + g = \mathrm{f}_0(\ell, e) + g = \ell + g + e\tilde{\mathrm{f}}(\ell, e) ,$$
$$r_2 = r_0(\ell, e) = m_0^{1/3} L^2 (1 - e \cos u_0(\ell, e)) .$$

where, as above, $e = e(L, G) = \sqrt{1 - (G/L)^2}$.

Notice that the positions (3.3.8) and (3.3.5) *define implicitly* m_0 *and hence* $m_0^{1/3}$ *as a (analytic) function of*[3.6] ε:

$$m_0(\varepsilon) = 1 - \varepsilon + \frac{2}{3}\varepsilon^2 - \frac{1}{3}\varepsilon^3 + \cdots ,$$
$$m_0(\varepsilon)^{1/3} = 1 - \frac{\varepsilon}{3} + \frac{1}{9}\varepsilon^2 - \frac{2}{81}\varepsilon^3 + \cdots .$$

Thus, introducing the functions

$$\begin{aligned} a_\varepsilon &:= a_\varepsilon(L) := m_0(\varepsilon)^{1/3} L^2 , \\ \rho_\varepsilon &:= \rho_\varepsilon(\ell, L, G) := a_\varepsilon(L) \Big(1 - e \cos u_0(\ell, e(L,G))\Big) , \\ \sigma &:= \sigma(\ell, L, G) := e(L,G)\, \tilde{\mathrm{f}}(\ell, e(L,G)) , \end{aligned}$$

we get

$$\tilde{H}_3 := \tilde{H}_2 \circ \hat{\phi}_{\rm lin} = -\frac{1}{2L^2} + \hat{T} - G + \varepsilon F_\varepsilon(\ell, g, L, G) ,$$

where

$$F_\varepsilon := \rho_\varepsilon \cos(\ell + g + \sigma) - \frac{1}{\sqrt{1 + \rho_\varepsilon^2 - 2\rho_\varepsilon \cos(\ell + g + \sigma)}} . \tag{3.3.10}$$

The variable $\hat{\tau}$ is cyclic (this is the reason for having introduced $\hat{\phi}_{\rm lin}$) and the linear constant of motion $\hat{T}$ can be dropped from $\tilde{H}_3$. The final form of *the Hamiltonian for the restricted, circular, planar, three-body-problem* is:

$$H_{\rm prc}(\ell, g, L, G; \varepsilon) := -\frac{1}{2L^2} - G + \varepsilon F_\varepsilon(\ell, g, L, G) ; \tag{3.3.11}$$

the phase space for the *inner* problem[3.7] may be taken to be

$$\mathbb{T}^2 \times \{(G, L) :\ 0 < G < L \ \text{ and } \ a_\varepsilon\,(1 + e) < 1\} ; \tag{3.3.12}$$

the symplectic form is the standard two-form $d\ell \wedge dL + dg \wedge dG$.

From the point of view of KAM theory, the integrable part of (3.3.11)

$$H_0(L, G) := H_{\rm prc}|_{\varepsilon=0} = -\frac{1}{2L^2} - G ,$$

is *iso-energetically non-degenerate* since

$$\det \begin{pmatrix} H_0'' & H_0' \\ H_0' & 0 \end{pmatrix} = \det \begin{pmatrix} -\frac{3}{L^4} & 0 & \frac{1}{L^3} \\ 0 & 0 & -1 \\ \frac{1}{L^3} & -1 & 0 \end{pmatrix} = \frac{3}{L^4} > 0 .$$

[3.6] In fact, from (3.3.8) one can invert the function $m_1 \to \varepsilon(m_1) = m_1/(1-m_1)^{2/3}$ and check that the inverse function $m_1(\varepsilon) = 1 - m_0(\varepsilon)$ has radius of convergence $(27/4)^{\frac{1}{3}} = 1.889881....$

[3.7] I.e., the problem where the mass-less body stays within the circle described by the primaries: $a_\varepsilon(1 + e)$ is the instantaneous aphelion (maximal distance from the Sun); in (3.3.12), a_ε and e have to be regarded as function of G and L.

3.4. The Sun-Jupiter-Asteroid problem

In order to apply the above theory to a physical model, we shall consider a restricted, circular, planar three-body problem extrapolated from the Solar system. As primary bodies, we take (obviously) the Sun (P_0) and Jupiter (P_1), and we shall approximate their motion by a *circular* one. In reality Jupiter is observed to revolve on an "osculating" ellipse of eccentricity

$$e_{\rm J} = 4.82 \cdot 10^{-2} \ ,$$

and, as we shall explain below, considering circular the orbit of Jupiter is the worst physical approximation in our model. The size of the perturbation parameter ε as defined above (see (3.3.8) and (3.3.5)) is given by[3.8]

$$\varepsilon_{_{\rm SJ}} := 0.954 \cdot 10^{-3} \simeq \frac{m_{\rm J}/(m_{\rm S}+m_{\rm J})}{\Big(m_{\rm S}/(m_{\rm S}+m_{\rm J})\Big)^{2/3}} \ . \tag{3.4.1}$$

As third body, assumed not to affect the motion of the primaries, we take an asteroid from the Asteroidal belt: this is an instance of *inner restricted problem* since the Asteroidal belt lies between the Sun and the trajectory of Jupiter. We shall consider asteroids whose observed osculating ellipses around the Sun are rather eccentric[3.9] ($e \simeq 0.2, 0.3$).

3.4.1. The Sun-Jupiter-Victoria model. To be specific we choose the asteroid *12 Victoria*[3.10] whose observed osculating data are the following

$$a_{\rm V} \simeq 0.449 \ , \qquad e_{\rm V} \simeq 0.220 \ , \qquad \imath_{\rm V} \simeq 1.961 \cdot 10^{-2} \ , \tag{3.4.2}$$

where: $a_{\rm V}$ denotes[3.11] the ratio between the observed semi-major axis of Victoria and that of Jupiter; $e_{\rm V}$ is the observed eccentricity of the osculating ellipse of Victoria and $\imath_{\rm V}$ is the relative inclination of the observed orbital planes of Victoria and Jupiter measured in degrees and normalized to one[3.12].

Clearly, in considering Sun-Jupiter-Victoria as a restricted, circular, planar three-body problem, we are making *a lot* of approximations disregarding: the eccentricity of Jupiter; the inclinations; the gravitational effects of all other bodies in the Solar system (most notably Mars and Saturn); the shape and extension of the bodies (in particular, asteroids are typically far from being spherical and, therefore, far from being well approximated by point masses); dissipative phenomena (tides, solar winds, Yarkovsky effect,...),...

A rough quantitative analysis shows, however, that *the worst approximation is actually to neglect Jupiter's eccentricity.*

Accordingly, we shall adopt the following "empirical criterion": *in considering the perturbation function F_ε in* (3.3.10) *we shall expand in*[3.13] *e and a neglecting the*

[3.8] $m_{\rm J}$:=mass of Jupiter $\simeq 1.9 \times 10^{27}$ Kg, $m_{\rm S}$:=mass of the Sun $\simeq 1.991 \times 10^{30}$ Kg; with our rescaling (see (3.3.5)), $m_0 = m_{\rm S}/(m_{\rm S}+m_{\rm J})$ and $m_1 = m_{\rm J}/(m_{\rm S}+m_{\rm J})$.

[3.9] There are various reasons to make this choice: we do not want to introduce extra "smallness" parameters besides the ratio of masses of the primaries; the mathematical model arising is "non-degenerate"; it is a case quite common in the observed data, ...

[3.10] The number refers to the standard classification of asteroidal objects; see, e.g., [**134**].

[3.11] Recall that in our units the orbital radius of the primaries is normalized to be 1.

[3.12] The observed inclination on the ecliptic of Jupiter is $1.305^{\rm o}$ while that of Victoria is $8.363^{\rm o}$ and the normalized relative inclination is $\imath_{\rm V} \simeq (8.363 - 1.305)/360$.

[3.13] F_ε is analytic for $|a| < 1$ and $|e| < 1$.

contributions of $O(e^3)$, $O(a^6)$, $O(e^2a^2)$ *and* $O(a^3e)$ since such terms in the case $e = e_{\rm V}$ and $a = a_{\rm V}$ are (definitely) smaller than $e_{\rm J}$:

$$e_{\rm V}^3 \simeq 1.06 \cdot 10^{-2} , \qquad a_{\rm V}^6 \simeq 8.19 \cdot 10^{-3} ,$$
$$e_{\rm V}^2 a_{\rm V}^2 \simeq 9.76 \cdot 10^{-3} , \qquad a_{\rm V}^3 e_{\rm V} \simeq 1.99 \cdot 10^{-2} .$$

Also, in view of (3.4.1), we shall neglect terms of order $O(\varepsilon)$ in F_ε, i.e., we shall replace F_ε by F_0. This amounts to replace $m_0^{1/3}$ with 1 and a_ε and ρ_ε with, respectively,

$$a_0 = L^2 , \qquad \rho_0 = a_0\Big(1 - e\cos u_0(\ell, e)\Big) , \qquad \Big(e = (1 - G^2/L^2)^{1/2}\Big) . \tag{3.4.3}$$

An easy computation (based on (3.2.29) and (3.2.30)) shows that, setting

$$F_0 = \sum_{j,k\geq 0} F_{jk}(\ell, g) e^j {a_0}^k ,$$

then the function $F_0^{\rm trunc}$ obtained from F_ε neglecting the above mentioned terms is given by:

$$\begin{aligned}
-F_0^{\rm trunc} &:= - \sum_{\substack{j<3, k<6\\ j+k<4\ \&\ j\geq 1}} F_{jk}(\ell, g) e^j {a_0}^k \\
&= 1 + \frac{{a_0}^2}{4} + \frac{9}{64}{a_0}^4 \\
&\quad -\frac{{a_0}^2\, e}{2} \cos\ell \\
&\quad +\Big(\frac{3}{8}{a_0}^3 + \frac{15}{64}{a_0}^5\Big)\cos(\ell + g) \\
&\quad -\frac{9}{4}{a_0}^2\, e\cos(\ell + 2g) \\
&\quad +\Big(\frac{3}{4}{a_0}^2 + \frac{5}{16}{a_0}^4\Big)\cos(2\ell + 2g) \\
&\quad +\frac{3}{4}{a_0}^2\, e\cos(3\ell + 2g) \\
&\quad +\Big(\frac{5}{8}{a_0}^3 + \frac{35}{128}{a_0}^5\Big)\cos(3\ell + 3g) \\
&\quad +\frac{35}{64}{a_0}^4\cos(4\ell + 4g) \\
&\quad +\frac{63}{128}{a_0}^5\cos(5\ell + 5g) .
\end{aligned}$$

We make a final addition: to (somehow) balance the fact that the lower harmonics are physically more important than the higher ones[3.14], we reintroduce in the lowest order harmonics the first discarded term. We are then led to the following one-parameter family of Hamiltonians:

$$\begin{aligned}
H_{\rm SJV}(\ell, g, L, G; \varepsilon) &:= -\frac{1}{2L^2} - G - \varepsilon P_{\rm SJV}(\ell, g, L, G) , \\
&=: H_0(L, G) + \varepsilon H_1(\ell, g, L, G) , \\
&\qquad (\ell, g) \in \mathbb{T}^2 , \quad 0 < G < L ,
\end{aligned} \tag{3.4.4}$$

[3.14] Think, for example, to averaging theory.

with

$$
\begin{aligned}
(3.4.5) \qquad P_{\rm SJV}(\ell,g,L,G) \;:=\; & 1+\frac{{a_0}^2}{4}+\frac{9}{64}\,{a_0}^4+\frac{3}{8}\,{a_0}^2e^2\\
& -\Big(\frac{1}{2}+\frac{9}{16}{a_0}^2\Big)\ {a_0}^2e\ \cos\ell\\
& +\Big(\frac{3}{8}\,{a_0}^3+\frac{15}{64}{a_0}^5\Big)\ \cos(\ell+g)\\
& -\Big(\frac{9}{4}+\frac{5}{4}{a_0}^2\Big)\ {a_0}^2e\ \cos(\ell+2g)\\
& +\Big(\frac{3}{4}\,{a_0}^2+\frac{5}{16}\,{a_0}^4\Big)\ \cos(2\,\ell+2\,g)\\
& +\frac{3}{4}\,{a_0}^2e\ \cos(3\,\ell+2\,g)\\
& +\Big(\frac{5}{8}\,{a_0}^3+\frac{35}{128}\,{a_0}^5\Big)\ \cos(3\,\ell+3\,g)\\
& +\frac{35}{64}\,{a_0}^4\ \cos(4\,\ell+4\,g)\\
& +\frac{63}{128}\,{a_0}^5\ \cos(5\,\ell+5\,g)\\
& \left(a_0=L^2\ ,\ e=\sqrt{1-\frac{G^2}{L^2}}\right)\ ,
\end{aligned}
$$

and *fixing the perturbation parameter at the value* $\varepsilon=\varepsilon_{\rm SJ}$, we obtain the *Sun-Jupiter-Victoria Hamiltonian*:

$$
\begin{aligned}
\overline{H}_{\rm SJV}(\ell,g,L,G) \;&:=\; -\frac{1}{2L^2}-G-\varepsilon_{\rm SJ}P_{\rm SJV}(\ell,g,L,G)\ ,\\
&=:\; H_0(L,G)+\varepsilon_{\rm SJ}H_1(\ell,g,L,G)\ .
\end{aligned}
$$

REMARK 3.4.1. (i) Note that

$$
{a_0}^2e^2=L^4-G^2L^2\ ,
$$

so that the eccentricity $e=e(L,G)$ appears in $P_{\rm SJV}$, only linearly and in front of the *odd* Fourier modes.

(ii) Observe, also, that

$$
{a_0}^2\frac{\partial E}{\partial L}=e^{-1}\ G^2L\ ,\qquad {a_0}^2\frac{\partial E}{\partial G}=-e^{-1}\ GL^2\ .
$$

Thus, in the gradient of the perturbation $P_{\rm SJV}$, the expressions which are not polynomial in (L,G) appear only through e and e^{-1}.

REMARK 3.4.2. (i) Of course, (many!) other inner asteroids with similar orbital characteristics could as well be considered: Victoria is just a sample (with an encouraging name). For example, our main stability theorem applies also to the asteroids Iris (7) and Renzia (1204). The orbital elements of such asteroids are given by:

$$
\begin{aligned}
&a_{\rm Iris}\simeq 0.459\ , && e_{\rm Iris}\simeq 0.230\ , && \imath_{\rm Iris}\simeq 1.172\cdot 10^{-2}\ ,\\
&a_{\rm Renzia}\simeq 0.435\ , && e_{\rm Renzia}\simeq 0.294\ , && \imath_{\rm Renzia}\simeq 1.603\cdot 10^{-3}\ .
\end{aligned}
$$

(ii) A numerical investigation of the validity of the model with Hamiltonian (3.4.4)-(3.4.5) has been performed in [**33**], where Laskar's frequency analysis ([**87**]) has been used to determine the critical break-down threshold of some invariant tori. The dynamics of the truncated Hamiltonian has been compared with the dynamics generated by the complete Hamiltonian function. The results show, in particular, that, in the case of Victoria, the break-down threshold of upper and lower bounding tori agrees for both models. A similar numerical analysis has been performed in order to evaluate the effect of the eccentricity of the orbit of Jupiter and of the relative inclination of the asteroid-Jupiter orbits and no relevant discrepancy has been found in both models.

3.4.2. The Sun-Jupiter-Victoria phase space. Let us now select the region of the phase space for the dynamical system (3.4.4), which may be considered more interesting from an astronomical point of view.

First of all, recall the observed osculating data for the asteroid Victoria (3.4.2), which are obtained computing the instantaneous osculating ellipses of the two-body system Sun-Asteroid (see [**134**]). In view of the approximations we have made and of the relation (3.4.3), we compute the corresponding "observed" value for the action variable L and hence, by the eccentricity relation in (3.4.3), the "observed" value of the action variable G:

$$\begin{aligned} \sqrt{a_{\rm V}} &\simeq 0.670 =: L_{\rm V} \ , \\ L_{\rm V} \sqrt{1 - e_{\rm V}^2} &\simeq 0.654 =: G_{\rm V} \ . \end{aligned}$$

Taking into account that the observed astronomical data are given in terms of "osculating" Keplerian ellipses, it seems reasonable to define the "osculating energy value" in terms of the Keplerian approximation. However, since the "secular" effects[3.15] are also certainly noticeable, we shall introduce them into the definition of the osculating energy value. Recalling (3.4.4), (3.4.5) and point (i) of Remark 3.4.1, we therefore define $E_{\rm V}^{(0)}$ and $E_{\rm V}^{(1)}$ as follows

$$\begin{aligned} H_0(L_{\rm V}, G_{\rm V}) &= -\frac{1}{2L_{\rm V}^2} - G_{\rm V} \simeq -1.768 =: E_{\rm V}^{(0)} \ , \\ \langle H_1(\cdot, L_{\rm V}, G_{\rm V}) \rangle &= -\Big(1 + \frac{L_{\rm V}^4}{4} + \frac{9}{64} L_{\rm V}^8 + \frac{3}{8} \left(L_{\rm V}^4 - G_{\rm V}^2 L_{\rm V}^2\right)\Big) \\ &\simeq -1.060 =: E_{\rm V}^{(1)} \ , \end{aligned}$$

$$E_{\rm V}(\varepsilon) := E_{\rm V}^{(0)} + \varepsilon E_{\rm V}^{(1)} \ . \tag{3.4.6}$$

The osculating energy level of the Sun-Jupiter-Victoria model is, then, defined as

$$\overline{E}_{\rm V} := E_{\rm V}(\varepsilon_{\rm SJ}) = E_{\rm V}^{(0)} + \varepsilon_{\rm SJ} E_{\rm V}^{(1)} \simeq -1.769 \ .$$

From now on, our theorems will be concerned with such one-parameter family of energy surfaces:

$$\mathcal{S}_{\varepsilon,{\rm V}} := H_{\rm SJV}^{-1}\big(E_{\rm V}(\varepsilon)\big) \ .$$

Next, we consider two invariant tori on $\mathcal{S}_{0,{\rm V}}$, which bound from above and below the observed value $L_{\rm V}$: we define

$$\tilde{L}_{\pm} = L_V \pm 0.001 \ .$$

[3.15] Roughly speaking, the effects of the average over the angles of the perturbation.

The corresponding frequencies are:

$$\tilde{\omega}_\pm := \frac{\partial H_0}{\partial(L,G)} = \Big(\frac{1}{\tilde{L}_\pm^3}, -1\Big) =: (\tilde{\alpha}_\pm, -1) \ .$$

Since we need Diophantine frequencies to study the KAM continuation of unperturbed tori, we proceed as follows. We compute the continued fraction representation up to the order 5 of the numbers $\tilde{\alpha}_\pm$ and, then, we modify the frequencies by adding a tail of all one's. In such a way we obtain two quadratic "noble" numbers $\alpha_\pm$ given by:

$$\begin{aligned} \alpha_- &:= [3;3,4,2,1^\infty] = 3.30976937631389... \\ \alpha_+ &:= [3;2,1,17,5,1^\infty] = 3.33955990647860... \ . \end{aligned} \tag{3.4.7}$$

We can now define the Diophantine frequencies

$$\omega_\pm := (\alpha_\pm, -1) \ . \tag{3.4.8}$$

The corresponding Diophantine constants are easily computed (compare § B.1, Appendix B):

$$\tau_\pm := \tau = 1 \ , \qquad \gamma_- := 7.224496 \cdot 10^{-3} \ , \qquad \gamma_+ := 3.324329 \cdot 10^{-2} \ .$$

We are interested in the KAM continuation of the following unperturbed tori, which lie on the energy level $H_0^{-1}(E_V^{(0)})$:

$$\mathcal{T}_0^\pm := \{(L_\pm, G_\pm)\} \times \mathbb{T}^2 \ , \tag{3.4.9}$$

with[3.16]

$$L_\pm := \frac{1}{\alpha_\pm^{1/3}} \ , \qquad G_\pm := -\frac{1}{2L_\pm^2} - E_V^{(0)} \ ,$$

and we shall prove the following statement, which shall be made precise in the next paragraph.

CLAIM 3.4.3. *The tori $\mathcal{T}_0^\pm$ (3.4.9) can be analytically continued for*[3.17] $|\varepsilon| \leq 10^{-3}$ *into invariant tori $\mathcal{T}_\varepsilon^\pm$ on the energy level $\mathcal{S}_{\varepsilon,\mathrm{V}} = H_{\mathrm{SJV}}^{-1}(E_\mathrm{V}(\varepsilon))$ keeping fixed the ratio of the frequencies. Since the system (3.4.4)-(3.4.5) is a two-degree-of-freedom, iso-energetically non-degenerate, the tori $\mathcal{T}_{\varepsilon_{\mathrm{SJ}}}^+$ and $\mathcal{T}_{\varepsilon_{\mathrm{SJ}}}^-$ are the boundary of an invariant region $\mathcal{J}$; such region contains the surface $(L_\mathrm{V}, G_\mathrm{V}) \times \mathbb{T}^2$, showing, in particular, that the motions*

$$(\ell(t), g(t), L(t), G(t)) := \phi^t_{H_{\mathrm{SJV}}}(\ell_0, g_0, L_\mathrm{V}, G_\mathrm{V})$$

belongs for any $t \in \mathbb{R}$ and any $(\ell_0, g_0) \in \mathbb{T}^2$ to the region $\mathcal{J}$. As a corollary the values of the perturbed integrals $L(t)$ and $G(t)$ stay close to their initial values L_V and G_V forever and the actual motion (in the mathematical model) is nearly elliptical with osculating orbital values close to the observed ones.

[3.16] The numerical values of $L_\pm$ and $G_\pm$ are

$$\begin{aligned} L_+ &= 0.671017866335225... \ , & G_+ &= 0.656922466367295... \ , \\ L_- &= 0.669016633073288... \ , & G_- &= 0.650269096020133... \ . \end{aligned} \tag{3.4.10}$$

[3.17] Recall from (3.4.1) that $\varepsilon_{\mathrm{SJ}} \simeq 0.954 \cdot 10^{-3} < 10^{-3}$.

CHAPTER 4

KAM Stability of the Sun-Jupiter-Victoria Problem

In this section we prove Claim 3.4.3, i.e., the stability of the Sun-Jupiter-Victoria problem modelled by a restricted, circular, planar three-body problem, as described in § 3.4.

The analysis in divided into four mains steps, which we proceed to briefly describe. All of these steps are "computer assisted", namely, they are implemented on a computer using the so-called "interval arithmetic" to keep track of the numerical errors introduced by the machine. Interval arithmetic is a standard technique, which allows to use computers in order to prove theorems; see Appendix C for more information.

1. Following § 2.8.2, we provide formulae for the explicit computation of the iso-energetic Lindstedt series (up to any order in ε) of KAM tori having Diophantine frequencies. We then compute the 12th-order truncation of the iso-energetic Lindstedt series for two tori $\mathcal{T}^{\pm}_{\varepsilon_{\text{SJ}}}$ (described in Claim 3.4.3); the energy level being fixed at the value $E_{\text{V}}(\varepsilon)$, see (3.4.6). Such 12th-order truncations of the iso-energetic Lindstedt series are taken as approximate tori[4.1]

$$(u^{(0)}, v^{(0)}, \omega^{(0)}) := (u^{(0)\pm}, v^{(0)\pm}, \omega^{(0)\pm}) \ . \tag{4.0.1}$$

By applying the KAM map $\mathcal{K}$ in (2.4.4) one can get better approximate tori

$$(u^{(i+1)\pm}, v^{(i+1)\pm}, \omega^{(i+1)\pm}) = \mathcal{K}(u^{(i)\pm}, v^{(i)\pm}, \omega^{(i)\pm}) \ . \tag{4.0.2}$$

In the next two steps, implementing with estimates this strategy, one shows that this is actually the case.

2. We provide formulae for estimating the input parameters of the KAM norm map $\widehat{\mathcal{K}}$ defined in Proposition 2.6.4 associated to the approximate tori defined in (4.0.1). The perturbation parameter ε is taken in the complex disk

$$\mathcal{E} := \{|\varepsilon| \leq \varepsilon_0\} := \{|\varepsilon| \leq 10^{-3}\} \ , \tag{4.0.3}$$

which contains the physical perturbation parameter ε_{SJ} of the Sun-Jupiter problem. Recall that in our choice of the norms it is taken the supremum over complex ε for $|\varepsilon| \leq \varepsilon_0$ so that, as already pointed out, the results holds uniformly for all $|\varepsilon| \leq 10^{-3}$; compare § 2.8.1.

[4.1]Often, to avoid heavy notations, we shall drop the suffix $\pm$, but keep in mind that, in actual computations, one has to treat separately the prolongation of the two tori $\mathcal{T}_0^{\pm}$.

3. We iterate the KAM map $\mathcal{K}$ a few times so as to obtain new approximate tori as in (4.0.2). The norms relative to these new approximate tori are controlled by means of the KAM norm map $\widehat{\mathcal{K}}$.

4. We finally apply the iso-energetic KAM theorem 2.7.1 (in the parameter dependent case described in § 2.8.1 with $\mathcal{E}$ as in (4.0.3)) to the two approximate tori

$$(u,v,\omega) = (u^+,v^+,\omega^+) := (u^{(3)+},v^{(3)+},\omega^{(3)+}) \ ,$$

and

$$(u,v,\omega) = (u^-,v^-,\omega^-) := (u^{(4)-},v^{(4)-},\omega^{(4)-}) \ ,$$

obtaining two invariant tori $(\tilde u^\pm,\tilde v^\pm,\tilde\omega^\pm)$ on the energy level $H_{\rm SJV}^{-1}(E_{\rm V})$. Claim 3.4.3 will then easily follow.

4.1. Iso-energetic Lindstedt series for the Sun-Jupiter-Asteroid problem and choice of the initial approximate tori $(u^{(0)\pm},v^{(0)\pm},\omega^{(0)\pm})$

Recall that the coefficients of the iso-energetic Lindstedt series, u_j, v_j, a_j, in (2.8.7) are determined recursively as follows (compare Proposition 2.8.3). The functions X_k, Y_k, Z_k in (2.8.8) are computed in terms of y_0, $\omega=\omega_0$ and $\{u_j,\ v_j,\ a_j\}$ with $1\le j\le k-1$ (for $k=1$, X_1, Y_1, Z_1 depend only upon y_0 and ω); then, (2.8.9), (2.8.10), (2.8.12) and (2.8.13) yield u_k, v_k and a_k.

In the Sun-Jupiter-Victoria model we have:

$$\begin{aligned}
&d=2\ , \qquad x=(x_1,x_2)=(\ell,g)\ , \qquad y=(y_1,y_2)=(L,G)\ ,\\
&H_0(y)=-\frac{1}{2y_1^2}-y_2\ , \qquad H_1(x,y)=-P_{\rm SJV}(x_1,x_2,y_1,y_2)\ ,\\
&y_0=(y_{01},y_{02})\ , \qquad \omega_0=(\omega_{01},\omega_{02})=\left(\frac{1}{y_{01}^3},-1\right)\ ,\\
&\mathcal{A}_0=\begin{pmatrix} -\frac{3}{y_{01}^4} & 0 & \frac{1}{y_{01}^3}\\ 0 & 0 & -1\\ \frac{1}{y_{01}^3} & -1 & 0\end{pmatrix}\ , \quad \mathcal{A}_0^{-1}=\begin{pmatrix} -\frac{y_{01}^4}{3} & -\frac{y_{01}}{3} & 0\\ -\frac{y_{01}}{3} & -\frac{1}{3y_{01}^2} & -1\\ 0 & -1 & 0\end{pmatrix}.
\end{aligned}\tag{4.1.1}$$

In view of § 3.4.2, the specific values to be considered are[4.2]

$$\begin{aligned}
&y_0=y^\pm=(y_1^\pm,y_2^\pm):=(L_\pm,G_\pm)\ , \qquad \omega_0=\omega_\pm=(\alpha_\pm,-1)\ ,\\
&E_0=E_{\rm V}^{(0)}=H_0(y^\pm)\ , \qquad E_1=E_{\rm V}^{(1)}\ , \qquad E_k=0\quad \forall k\ge 2\ .
\end{aligned}\tag{4.1.2}$$

We shall treat simultaneously the two cases corresponding to the two tori $\mathcal{T}_\varepsilon^\pm$, often omitting in the notation the suffix $^\pm$.

In order, to compute explicitly (and effectively) the expansions appearing in the definitions of $X_k=X_k^\pm$, $Y_k=Y_k^\pm$, $Z_k=Z_k^\pm$ (see (2.8.8)), it is helpful the following simple

---

[4.2]See, in particular, (2.8.5), (2.8.6), (3.4.6) and (3.4.8).

LEMMA 4.1.1. *For $p, j \in \mathbb{Z}_+$, denote by $\mathcal{I}_j^p$ the following set of indices*

$$\mathcal{I}_j^p := \Big\{\alpha^{(1)}, ..., \alpha^{(p)} \in \mathbb{N}^j : \sum_{\substack{1\le \ell\le j\\ 1\le q\le p}} \ell\ \alpha_\ell^{(q)} = j\Big\} .$$

Let

$$f = \sum_{r\in\mathbb{N}^d} f_r\, (y-y_0)^r$$

be a convergent power series in a neighborhood of $y_0 \in \mathbb{C}^d$; let

$$\varepsilon \in \mathbb{C} \to \sum_{j=0}^{\infty} w_j \varepsilon^j \in \mathbb{C}^d$$

be convergent power series such that $w_0 = y_0$; $w_j = (w_{j1}, ..., w_{jd})$. Then

$$f\Big(\sum_{j=0}^{\infty} w_j\varepsilon^j\Big) = f(y_0)$$

$$+\sum_{j=1}^{\infty}\left(\sum_{\mathcal{I}_j^d} f_{(|\alpha^{(1)}|_1,...,|\alpha^{(d)}|_1)} \frac{|\alpha^{(1)}|_1!\cdots|\alpha^{(d)}|_1!}{\alpha^{(1)}!\cdots\alpha^{(d)}!} \prod_{\substack{1\le\ell\le j\\ 1\le q\le d}} w_{\ell q}^{\alpha_\ell^{(q)}}\right)\varepsilon^j .$$

The proof is standard[4.3]. □

Another useful (from the computational viewpoint) simple result is:

LEMMA 4.1.2. *Let $\varepsilon \in \mathbb{C} \to \sum_{k=0}^{\infty} w_k\varepsilon^k \in \mathbb{C}$ be a complex power series. Then,*

$$\exp\Big(\sum_{k=0}^{\infty} w_k\varepsilon^k\Big) = \sum_{k=0}^{\infty} b_k\varepsilon^k , \tag{4.1.3}$$

with $b_0 = \exp(w_0)$ and, for $k \ge 1$,

$$b_k = \frac{1}{k}\sum_{j=1}^{k} j b_{k-j} w_j = \frac{1}{k}\sum_{j=0}^{k-1}(k-j)\, b_j w_{k-j} . \tag{4.1.4}$$

To check (4.1.4), differentiate (4.1.3) with respect to ε, use (4.1.3) to eliminate the exponential and equate coefficients of the power series. □

Before proceeding, we recall in the next remark the form of the perturbation function $H_1 = -P_{\rm SJV}$.

REMARK 4.1.3. From § 3.4.1 it follows that H_1 and its derivatives are of the form

$$\sum_{n\in\mathcal{N}} f_n(y) c_n(x) ,$$

where $\mathcal{N}$ is a finite set in $\mathbb{Z}^2$, f_n are real-analytic functions in a neighborhood of y_0,

$$f_n(y) = \sum_{r\in\mathbb{N}^2} f_n^{(r)}\, (y-y_0)^r , \tag{4.1.5}$$

[4.3] Recall that $\Big(\sum_{\ell=1}^{n} x_\ell\Big)^j = \sum_{\alpha\in\mathbb{N}^n:\ |\alpha|_1=j} \frac{j!}{\alpha!}\, x^\alpha$.

and the c_n's are either all $\cos(n\cdot x)$ (if the number of x-derivatives of H_1 is even) or all $\sin(n\cdot x)$ (if the number of x-derivatives of H_1 is odd). Furthermore, (recall also Remark 3.4.1) the Fourier coefficients f_n are of the form

$$c\, y_1^j y_2^k e(y)^\ell\ , \quad e(y) := \sqrt{1-(y_2/y_1)^2}\ , \quad \Big(c\in\mathbb{R}\ ,\ j,\ell\in\mathbb{Z}\ ,\ k\in\mathbb{N}\Big)\ . \tag{4.1.6}$$

In fact, for the first order derivatives of H_1 the above exponents j and k assume only non-negative values while ℓ assumes the values $-1, 0, 1$. The computation of the coefficients $f_n^{(r)}$ is seen to be a simple exercise, which can be implemented, in a fast way, on a computer.

Let us turn to the "explicit" computation of X_k Y_k and Z_k in (2.8.8).

Let us set $(u_0, v_0, a_0) = (0, y_0, 0)$ and assume to know the coefficients of the iso-energetic Lindstedt series

$$(u_j, v_j, a_j) = (u_j^\pm, v_j^\pm, a_j^\pm)\ , \qquad j\le k-1\ ,$$

for some $k\ge 1$.

In view of Lemma 4.1.2, for $0\le h\le k-1$, we find that[4.4]

$$\begin{aligned}
\left[\cos\Big(n\cdot\theta+\sum_{j=1}^{k-1} n\cdot u_j\,\varepsilon^j\Big)\right]_h &= \frac{b_n^{(h)}(\theta)+b_{-n}^{(h)}(\theta)}{2}\ ,\\
\left[\sin\Big(n\cdot\theta+\sum_{j=1}^{k-1} n\cdot u_j\,\varepsilon^j\Big)\right]_h &= \frac{b_n^{(h)}(\theta)-b_{-n}^{(h)}(\theta)}{2i}\ ,
\end{aligned} \tag{4.1.7}$$

where the $b_n^{(h)}$ are recursively defined as

$$\begin{aligned}
b_n^{(0)}(\theta) &= \exp(in\cdot\theta)\ ,\\
b_n^{(h)}(\theta) &= \frac{1}{h}\sum_{j=1}^{h} j b_n^{(h-j)}\,(n\cdot u_j) = \frac{1}{h}\sum_{j=0}^{h-1}(h-j)\,b_n^{(j)}(n\cdot u_{h-j})\ .
\end{aligned}$$

Thus, by Lemma 4.1.1 and (4.1.7), one finds, for $h=k-1$ or $h=k$,

$$\begin{aligned}
&\left[\sum_{n\in\mathcal{N}} f_n\Big(y_0+\sum_{j=1}^{k-1}\varepsilon^j v_j\Big)\, c_n\Big(\theta+\sum_{j=1}^{k-1}\varepsilon^j u_j\Big)\right]_h\\
&= \sum_{\substack{0\le\ell\le h\\ n\in\mathcal{N}}}\ \sum_{\alpha^{(1)},\alpha^{(2)}\in\mathcal{I}_\ell^2} \frac{b_n^{(h-\ell)}+b_{-n}^{(h-\ell)}}{2}\, f_n^{(|\alpha^{(1)}|_1,|\alpha^{(2)}|_1)}\,\frac{|\alpha^{(1)}|_1!\,|\alpha^{(2)}|_1!}{\alpha^{(1)}!\,\alpha^{(2)}!}\\
&\qquad\cdot\prod_{\substack{1\le\ell'\le\ell\\ 1\le q\le 2}}\Big(f_n^{(\ell' q)}\Big)^{\alpha_{\ell'}^{(q)}}\ ,
\end{aligned}$$

[4.4]Recall that $\Big[\cdot\Big]_j := \frac{1}{j!}\frac{d^j}{d\varepsilon^j}\Big|_{\varepsilon=0}$.

when the $c_n(x)$'s are $\cos(n\cdot x)$'s, and

$$\left[\sum_{n\in\mathcal{N}} f_n\Big(y_0+\sum_{j=1}^{k-1}\varepsilon^j v_j\Big)\; c_n\Big(\theta+\sum_{j=1}^{k-1}\varepsilon^j u_j\Big)\right]_h$$
$$=\sum_{\substack{0\le\ell\le h\\ n\in\mathcal{N}}}\;\sum_{\alpha^{(1)},\alpha^{(2)}\in\mathcal{I}_\ell^2}\frac{b_n^{(h-\ell)}-b_{-n}^{(h-\ell)}}{2i}\; f_n^{(|\alpha^{(1)}|_1,|\alpha^{(2)}|_1)}\;\frac{|\alpha^{(1)}|_1!\,|\alpha^{(2)}|_1!}{\alpha^{(1)}!\,\alpha^{(2)}!}$$
$$\cdot\prod_{\substack{1\le\ell'\le\ell\\1\le q\le 2}}\Big(a_n^{(\ell' q)}\Big)^{\alpha_{\ell'}^{(q)}}\;,$$

when the $c_n(x)$'s are $\sin(n\cdot x)$'s.

At this point, it is simple to compute the Fourier coefficients of $X_k^\pm$ and $Y_k^\pm$ and to compute, via (2.8.9), (2.8.10), (2.8.12) and (2.8.13), the (Fourier coefficients of the) functions $u_k^\pm$, $v_k^\pm$ and the number $a_k^\pm$.

Implementing the above formulae on a computer machine, keeping track of numerical errors by interval arithmetic, one can, now, compute the iso-energetic Lindstedt series for the Sun-Jupiter-Victoria model up to order $k=12$.

In fact, in view of the particular symmetry of the interaction in the three-body-problem ($P_{\rm SJV}$ is even in x), it turns out that the u_k's are *odd* in θ and the v_k's are *even* in θ. Therefore the u_k's and v_k's have a representation of the form

$$\begin{aligned} u_k^\pm &= \sum_{n\in\mathcal{N}_k^u} u_n^{k\pm}\sin(n\cdot\theta)\;,\\ v_k^\pm &= \sum_{n\in\mathcal{N}_k^v} v_n^{k\pm}\cos(n\cdot\theta)\;, \end{aligned}\tag{4.1.8}$$

with $u_n^{k\pm}$ and $v_n^{k\pm}$ in $\mathbb{R}^2$ and where $\mathcal{N}_k^u$ and $\mathcal{N}_k^v$ are suitable *finite* subset of $\mathbb{Z}^2$. In Table 1 we report the number of non-vanishing Fourier coefficients, denoted by $N_{\rm Four}(\cdot)$, of each component of u_k and v_k for $k\le 12$; the cardinality of $\mathcal{N}_k^u$ and $\mathcal{N}_k^v$ coincide, respectively, with $N_{\rm Four}(u_{k1})$ and $N_{\rm Four}(v_{k1})$.

We now define the approximate torus $(u^{(0)},v^{(0)},\omega^{(0)})=(u^{(0)\pm},v^{(0)\pm},\omega^{(0)\pm})$ as the $12^{\rm th}$-order truncation of the iso-energetic Lindstedt series:

$$(u^{(0)},v^{(0)},\omega^{(0)}):=\left(\sum_{k=1}^{12}\varepsilon^k u_k^\pm,\, y^\pm+\sum_{k=1}^{12}\varepsilon^k v_k^\pm,\,(1+a_\pm)\omega_\pm\right)\tag{4.1.9}$$

with

$$a_\pm:=\sum_{k=1}^{12}a_k^\pm\varepsilon^k\;.\tag{4.1.10}$$

In particular, following the above formulae and using interval arithmetic on a computer, we find that

$$a_+=-2.217750...\cdot 10^{-4}\;,\qquad a_-=-2.255270...\cdot 10^{-4}\;.\tag{4.1.11}$$

TABLE 1. Number of Fourier modes of the Lindstedt series k^{th}-coefficient

k	$N_{\text{Four}}(u_{k1})$	$N_{\text{Four}}(u_{k2})$	$N_{\text{Four}}(v_{k1})$	$N_{\text{Four}}(v_{k2})$
1	8	3	9	8
2	38	28	40	39
3	104	89	107	106
4	220	200	224	223
5	400	375	405	404
6	658	628	664	663
7	1008	973	1015	1014
8	1464	1424	1472	1471
9	2040	1995	2049	2048
10	2750	2700	2760	2759
11	3608	3553	3619	3618
12	4628	4568	4640	4639

4.2. Evaluation of the input parameters of the KAM norm map associated to the approximate tori $\left(u^{(0)\pm}, v^{(0)\pm}, \omega^{(0)\pm}\right)$

4.2.1. Choice of analyticity radii and auxiliary parameters. To evaluate the KAM norm map $\widehat{\mathcal{K}}$ defined in Proposition 2.6.4 (see also § 2.8.1 for the parameter-dependent case), we have to fix the parameters $\bar{\xi}$, r (which measure the domain where H_{SJV} is considered), $\xi^{(0)}$ (which measures the domain where the approximate torus is considered), κ (an auxiliary prefixed number in $(1,2]$), ρ (a norm weight) and ε_0 (which measures the domain where the parameter ε is allowed to vary). It is convenient to make the following choices[4.5]:

$$\bar{\xi}^+ := 0.103782\ldots\ , \qquad r^+ := 4.481700\ldots \cdot 10^{-4}\ ,$$
$$\bar{\xi}^- := 0.104159\ldots\ , \qquad r^- := 4.304517\ldots \cdot 10^{-4}\ ,$$
$$\varepsilon_0 := 10^{-3}\ , \qquad \xi^{(0)} := 0.1\ ,$$
$$\kappa := 1.01\ , \qquad \rho := 0.2\ .$$

4.2.2. Evaluation of $E_{p,q}$. Recall that $E_{p,q}$ are upper bounds on the $(p+q)$-tensor of derivatives $\partial_x^p \partial_y^q H$ (see (2.7.1)). Recalling (2.5.8), we shall let $E_{p,q}$ be positive numbers such that

$$\sqrt{\sum_{j_1,\ldots,j_q} \left(\|\partial^q_{y_{j_1}\cdots y_{j_q}} H_0\|_r \ + \ \varepsilon_0 \ \|\partial^q_{y_{j_1}\cdots y_{j_q}} H_1\|_{\bar{\xi},r} \right)^2} \leq E_{0,q}\ ,$$

$$\varepsilon_0 \sqrt{\sum_{\substack{i_1,\ldots,i_p \\ j_1,\ldots,j_q}} \|\partial^p_{x_{i_1}\cdots x_{i_p}} \partial^q_{y_{j_1}\cdots y_{j_q}} H_1\|^2_{\bar{\xi},r}} \leq E_{p,q}\ , \qquad \text{if } p \neq 0\ ,$$

where the indices i_k and j_ℓ take the values 1 and 2.

The various terms appearing in the derivatives of H_i can be estimated making use of the majorizations discussed in § 2.5.6 (see, in particular (2.5.15), (2.5.14)), as

[4.5] Such choices have been made essentially by a "trial and error" method, trying to optimize the various algorithms and programs involved.

follows. If $k \in \mathbb{N}$ and $i = 1, 2$, we have[4.6]

$$\|y_i^k\|_r \le (y_{0i} + r)^k \ , \qquad \|y_i^{-k}\|_r \le (y_{0i} - r)^{-k} \ . \tag{4.2.1}$$

Positive and negative powers of $e(y)$ may be bounded by ($k \ge 1$):

(4.2.2)

$$\|e(y)^k\|_r \le \left(2 - \sqrt{1 - \left(\frac{y_{02}+r}{y_{01}-r}\right)^2}\right)^k \ , \qquad \|e(y)^{-k}\|_r \le \left(1 - \left(\frac{y_{02}+r}{y_{01}-r}\right)^2\right)^{-\frac{k}{2}} \ .$$

The function H_0 and its derivatives are of the form $cy_1^{-j} + y_2^k$ with $c \in \mathbb{R}$, $j, k \in \mathbb{N}$ and such terms can be bounded, using (4.2.1), by

$$\|cy_1^{-j} + y_2^k\|_r \le |c| \ (y_{01} - r)^{-j} + (y_{02} + r)^k \ .$$

For example, if $q \ge 2$, then

$$\begin{aligned} \sqrt{\sum_{j_1,\dots,j_q} \|\partial^q_{y_{j_1}\cdots y_{j_q}} H_0\|_r^2} &= \|\partial^q_{y_1} H_0\|_r = \frac{(q+1)!}{2} \left\|y_{01}^{-(q+2)}\right\|_r \\ &\le \frac{(q+1)!}{2} \, \frac{1}{(L_\pm - r)^{q+2}} \ , \qquad (q \ge 2) \ . \end{aligned}$$

The function H_1 and its derivatives, as pointed out in Remark 4.1.3, are of the form $\sum_{n\in\mathcal{N}} a_n(y)c_n(x)$, with $a_n(y)$ as in (4.1.6). Since[4.7]

$$\Big\| \sum_{n\in\mathcal{N}} a_n(y)c_n(x) \Big\|_{r,\bar\xi} \le \sum_{n\in\mathcal{N}} \|a_n(y)\|_r \exp(|n|\bar\xi) \ ,$$

we see that H_1 and its derivatives may be estimated straightforwardly using (4.2.1) and (4.2.2).

Implementing this discussion one obtains the following numerical values:

$$\begin{aligned} &y_{01}^+ + r^+ \le 0.669464... \ , \qquad y_{02}^+ + r^+ \le 0.650717... \ , \\ &(y_{01}^+ - r^+)^{-1} \le 1.495733... \ , \qquad (y_{02}^+ - r)^{-1} \le 1.538885... \ , \\ &2 - \sqrt{1 - \left(\frac{y_{02}^+ + r^+}{y_{01}^+ - r^+}\right)^2} \le 1.770460... \ , \\ &\left(1 - \left(\frac{y_{02}^+ + r^+}{y_{01}^+ - r^+}\right)^2\right)^{-\frac{1}{2}} \le 4.356550... \ , \end{aligned}$$

$$\begin{aligned} &y_{01}^- + r^- \le 0.671448... \ , \qquad y_{02}^- + r^- \le 0.657352... \ , \\ &(y_{01}^- - r^-)^{-1} \le 1.491229... \ , \qquad (y_{02}^- - r^-)^{-1} \le 1.523247... \ , \\ &2 - \sqrt{1 - \left(\frac{y_{02}^- + r^-}{y_{01}^- - r^-}\right)^2} \le 1.802308... \ , \\ &\left(1 - \left(\frac{y_{02}^- + r^-}{y_{01}^- - r^-}\right)^2\right)^{-\frac{1}{2}} \le 5.058391... \ , \end{aligned}$$

[4.6] Recall (see (3.4.10)) that $y_0 = (L_\pm, G_\pm) \in \mathbb{R}^2_+$ and that $r^\pm \ll y_{0i}^\pm$.

[4.7] Recall (2.5.11).

and

$$\begin{aligned}
E^+_{1,0} &= 5.373428...\cdot 10^{-3}\ , & E^+_{0,1} &= 3.515464...\ ,\\
E^+_{1,1} &= 5.685744...\cdot 10^{-2}\ , & E^+_{0,2} &= 15.384979...\ ,\\
E^+_{1,2} &= 2.211666...\ , & E^+_{2,1} &= 0.413277...\ ,\\
E^+_{0,3} &= 119.267702...\ , & E^+_{3,0} &= 5.865020...\cdot 10^{-4}\ ,
\end{aligned}$$

$$\begin{aligned}
E^-_{1,0} &= 5.527575...\cdot 10^{-3}\ , & E^-_{0,1} &= 3.488445...\ ,\\
E^-_{1,1} &= 6.190221...\cdot 10^{-2}\ , & E^-_{0,2} &= 15.335556...\ ,\\
E^-_{1,2} &= 3.170738...\ , & E^-_{2,1} &= 0.446508...\ ,\\
E^-_{0,3} &= 159.060236...\ , & E^-_{3,0} &= 6.050803...\cdot 10^{-4}\ .
\end{aligned}$$

4.2.3. Evaluation of $U^{(0)\pm}$, $V^{(0)\pm}$, $\tilde V^{(0)\pm}$, $M^{(0)\pm}$, $\overline{M}^{(0)\pm}$. Recall the definitions of U, V, $\tilde V$, M and $\overline{M}$ in (2.6.2). In view of the representation (4.1.8), we have that $\sup_{\mathbb{T}^2_{\xi^{(0)}}} |\operatorname{Im} u^\pm_k|$ and $\|v^\pm_k\|_{\xi^{(0)}}$ can be bounded by positive numbers $U^{(0)\pm}_k$ and $V^{(0)\pm}_k$ as follows[4.8]

$$\begin{aligned}
\sup_{\mathbb{T}^2_{\xi^{(0)}}} |\operatorname{Im} u^\pm_k| &\le \sum_{n\in\mathcal{N}^u_k} |u^{k\pm}_n|\sinh(|n|_1\xi^{(0)}) \le U^{(0)\pm}_k\ ,\\
\|v^\pm_{k1}\|_{\xi^{(0)}} &\le \sum_{n\in\mathcal{N}^v_k} |v^{k\pm}_{n1}|\exp(|n|\xi^{(0)}) \le V^{(0)\pm}_{k1}\ ,\\
\|v^\pm_{k2}\|_{\xi^{(0)}} &\le \sum_{n\in\mathcal{N}^v_k} |v^{k\pm}_{n2}|\exp(|n|\xi^{(0)}) \le V^{(0)\pm}_{k2}\ ,\\
\|v^\pm_{k}\|_{\xi^{(0)}} &\le \sum_{n\in\mathcal{N}^v_k} |v^{k\pm}_{n}|\exp(|n|\xi^{(0)}) \le V^{(0)\pm}_{k}\ .
\end{aligned}\tag{4.2.3}$$

Then, $U^{(0)\pm}$, $V^{(0)\pm}$ can be taken to be numbers such that[4.9]

$$\begin{aligned}
&\sup_{\mathbb{T}^d_{\xi^{(0)}},\mathcal{E}} |\operatorname{Im} u^{(0)\pm}| \le \sum_{k=1}^{12} \varepsilon_0^k U^{(0)\pm}_k \le U^{(0)\pm}\ ,\\
&\|v^{(0)\pm}_1 - y^\pm_{01}\|_{\xi^{(0)},\mathcal{E}} \le \sum_{k=1}^{12} \varepsilon_0^k V^{(0)\pm}_{k1} \le \check V^{(0)\pm}_1\ ,\\
&\|v^{(0)\pm}_2 - y^\pm_{02}\|_{\xi^{(0)},\mathcal{E}} \le \sum_{k=1}^{12} \varepsilon_0^k V^{(0)\pm}_{k2} \le \check V^{(0)\pm}_2\ ,\\
&\|v^{(0)\pm} - y^\pm_{0}\|_{\xi^{(0)},\mathcal{E}} \le \sum_{k=1}^{12} \varepsilon_0^k V^{(0)\pm}_{k} \le \check V^{(0)\pm}\ ,\\
&\|v^{(0)\pm}_1\|_{\xi^{(0)},\mathcal{E}} \le |y^\pm_{01}| + \|v^{(0)\pm}_1 - y^\pm_{01}\|_{\xi^{(0)},\mathcal{E}} \le |y^\pm_{01}| + \check V^{(0)\pm}_1 = V^{(0)\pm}_1\ ,\\
&\|v^{(0)\pm}_2\|_{\xi^{(0)},\mathcal{E}} \le |y^\pm_{02}| + \check V^{(0)\pm}_2 = V^{(0)\pm}_2\ ,\\
&\|v^{(0)\pm}\|_{\xi^{(0)},\mathcal{E}} \le |y^\pm_{0}| + \|v^{(0)\pm} - y^\pm_{0}\|_{\xi^{(0)},\mathcal{E}} \le |y^\pm_{0}| + \check V^{(0)\pm} = V^{(0)\pm}\ .
\end{aligned}\tag{4.2.4}$$

[4.8]Observe that for $n\in\mathbb{Z}^d$ and $\theta\in\mathbb{T}^d_\xi$ one has $\big|\operatorname{Im}\,\sin(n\cdot\theta)\big| \le \sinh(|n|_1\xi)$.

[4.9]Recall that $\mathcal{E} := \{\varepsilon\in\mathbb{C}:\ |\varepsilon|\le\varepsilon_0\}$, $\varepsilon_0 := 10^{-3}$.

Analogously, $\|\partial_\theta u^\pm\|_{\xi^{(0)}}$ and $\|\partial_\theta v^\pm\|_{\xi^{(0)}}$ can be bounded by positive numbers $\tilde{U}^{(0)\pm}$ and $\tilde{V}^{(0)\pm}$, such that

$$\begin{aligned}
\|\partial_\theta u_k^\pm\|_{\xi^{(0)}} &\le \sqrt{\sum_{i,j=1}^{2}\Big(\sum_{k=1}^{12}\varepsilon_0^k\sum_{n\in\mathcal{N}_k^u}|n_i|\,|u_{nj}^{k\pm}|\exp(|n|\xi^{(0)})\Big)^2} \le \tilde{U}^{(0)\pm}\ ,\\
\|\partial_\theta v_k^\pm\|_{\xi^{(0)}} &\le \sqrt{\sum_{i,j=1}^{2}\Big(\sum_{k=1}^{12}\varepsilon_0^k\sum_{n\in\mathcal{N}_k^v}|n_i|\,|v_{nj}^{k\pm}|\exp(|n|\xi^{(0)})\Big)^2} \le \tilde{V}^{(0)\pm}\ .
\end{aligned}$$

We shall also need bounds on $|\partial_\theta u_k^\pm|$ and $|v^{(0)\pm}-y_0^\pm|$ for *real* θ, which are simply given by

$$\|\partial_\theta u_k^\pm\|_{0,\varepsilon} \le \sqrt{\sum_{i,j=1}^{2}\Big(\sum_{k=1}^{12}\varepsilon_0^k\sum_{n\in\mathcal{N}_k^u}|n_i|\,|u_{nj}^{k\pm}|\Big)^2} \le \tilde{U}_{\text{real}}^{(0)\pm}\ ,$$

$$\|v^{(0)\pm}-y_0^\pm\|_{0,\varepsilon} \le \sum_{k=1}^{12}\varepsilon_0^k\sum_{n\in\mathcal{N}_k^v}|v_n^{k\pm}| \le \check{V}_{\text{real}}^{(0)\pm}\ . \tag{4.2.5}$$

Finally, as $M^{(0)\pm}$, $\overline{M}^{(0)\pm}$, we can take positive numbers such that

$$1+\tilde{U}^{(0)\pm} \le M^{(0)\pm}\ , \qquad (1-\tilde{U}^{(0)\pm})^{-1} \le \overline{M}^{(0)\pm}\ .$$

Implementing this discussion one obtains the following numerical values:

$$\begin{aligned}
U^{(0)+} &= 3.772624...\cdot 10^{-3}\ , & V^{(0)+} &= 0.933497...\ ,\\
\check{V}_1^{(0)+} &= 3.144031...\cdot 10^{-4}\ , & \check{V}_2^{(0)+} &= 4.084723...\cdot 10^{-4},\\
\tilde{V}^{(0)+} &= 1.292660...\cdot 10^{-3}\ , & \check{V}_{\text{real}}^{(0)+} &= 3.130263...\cdot 10^{-4}\ ,\\
\tilde{U}^{(0)+} &= 1.579746...\cdot 10^{-2}\ , & \tilde{U}_{\text{real}}^{(0)+} &= 1.302041...\cdot 10^{-2}\ ,\\
M^{(0)+} &= 1.015797...\ , & \overline{M}^{(0)+} &= 1.016051...\ ,
\end{aligned}$$

$$\begin{aligned}
U^{(0)-} &= 4.149873...\cdot 10^{-3}\ , & V^{(0)-} &= 0.939554...\ ,\\
\check{V}_1^{(0)-} &= 3.052917...\cdot 10^{-4}\ , & \check{V}_2^{(0)-} &= 3.900402...\cdot 10^{-4},\\
\tilde{V}^{(0)-} &= 1.256447...\cdot 10^{-3}\ , & \check{V}_{\text{real}}^{(0)-} &= 3.014611...\cdot 10^{-4}\ ,\\
\tilde{U}^{(0)-} &= 1.737659...\cdot 10^{-2}\ , & \tilde{U}_{\text{real}}^{(0)-} &= 1.429971...\cdot 10^{-2}\ ,\\
M^{(0)-} &= 1.017376...\ , & \overline{M}^{(0)-} &= 1.017683...\ .
\end{aligned}$$

In particular we have (see, also, (4.2.4))

$$U^{(0)+} = 3.772624... \cdot 10^{-3} < 3.782624... \cdot 10^{-3} = \bar{\xi}^{(0)} - \xi^{(0)} ,$$

$$\sup_{\mathbb{T}^2_{\xi^{(0)}}} |v^{(0)+}(\theta) - y_0^+|_\infty \leq \max\{\check{V}_1^{(0)+}, \check{V}_2^{(0)+}\} = 4.084723... \cdot 10^{-4}$$

$$< 4.481700... \cdot 10^{-4} = r ,$$

$$U^{(0)-} = 4.149873... \cdot 10^{-3} < 4.159873... \cdot 10^{-3} = \bar{\xi} - \xi^{(0)} ,$$

$$\sup_{\mathbb{T}^2_{\xi^{(0)}}} |v^{(0)-}(\theta) - y_0^-|_\infty \leq \max\{\check{V}_1^{(0)-}, \check{V}_2^{(0)-}\} = 3.900402... \cdot 10^{-4}$$

$$< 4.304517... \cdot 10^{-4} = r ,$$

showing that (2.6.3) is verified for our two approximate tori.

4.2.4. Evaluation of $\overline{A}^{(0)\pm}$. Recall that $\overline{A}^{(0)\pm}$ is an upper bound on the norm of the (constant) matrix $(\mathcal{A}^{(0)\pm})^{-1}$, where $\mathcal{A}^{(0)\pm}$ is the matrix $\mathcal{A}$ in[4.10] (2.4.3) with (u, v, ω) replaced by $(u^{(0)\pm}, v^{(0)\pm}, \omega^{(0)\pm})$. Dropping the suffix $\pm$ we define the matrix $\hat{\mathcal{A}}_0$ as follows[4.11]

$$\hat{\mathcal{A}}_0 := \begin{pmatrix} \partial_y^2 H_0(y_0) & -\frac{\omega_0}{\rho} \\ \frac{\omega_0}{\rho} & 0 \end{pmatrix} = \begin{pmatrix} -\frac{3}{y_{01}^4} & 0 & -\frac{1}{\rho y_{01}^3} \\ 0 & 0 & \frac{1}{\rho} \\ \frac{1}{\rho y_{01}^3} & -\frac{1}{\rho} & 0 \end{pmatrix} ,$$

$$\hat{\mathcal{A}}_0^{-1} = \begin{pmatrix} -\frac{y_{01}^4}{3} & -\frac{y_{01}}{3} & 0 \\ -\frac{y_{01}}{3} & -\frac{1}{3y_{01}^2} & -\rho \\ 0 & \rho & 0 \end{pmatrix} .$$

Define, also, $\mathcal{A}_1$ by letting

$$\mathcal{A} = \mathcal{A}_0 + \mathcal{A}_1 , \qquad \mathcal{A}_1 := \begin{pmatrix} \langle \mathcal{T} \rangle - \partial_y^2 H_0(y_0) & -\left(\langle \chi \rangle - \frac{\omega_0}{\rho}\right) \\ \chi(0) - \frac{\omega_0}{\rho} & 0 \end{pmatrix} . \tag{4.2.6}$$

We, then, will estimate

$$|\mathcal{A}^{-1}|_\mathcal{E} = |(I + \mathcal{A}_0^{-1}\mathcal{A}_1)^{-1}\mathcal{A}_0^{-1}|_\mathcal{E} \leq |\mathcal{A}_0^{-1}|\,(1 - |\mathcal{A}_0^{-1}|\,|\mathcal{A}_1|_\mathcal{E})^{-1} . \tag{4.2.7}$$

A bound on $\mathcal{A}_0$ is immediately gotten using (2.5.2):

$$|\mathcal{A}_0^{-1}| \leq \sqrt{\frac{L_\pm^8}{9} + \frac{2L_\pm^2}{9} + \frac{1}{9L_\pm^4} + 2\rho^2} . \tag{4.2.8}$$

To give an upper bound on $|\mathcal{A}_1|_\mathcal{E}$ we proceed as follows. Define the (2×2)-matrices $\mathcal{B}$ and $\mathcal{C}$ and the vector b by letting

$$\mathcal{M}^{-1} = I + \mathcal{B} , \qquad \partial_y^2 H(\theta + u^{(0)}, v^{(0)}) = \partial_y^2 H_0(y_0) + \mathcal{C} , \qquad \chi = \frac{\omega_0 + b}{\rho} . \tag{4.2.9}$$

[4.10] We recall here the definition of $\mathcal{A}$ in (2.3.35):

$$\mathcal{A} := \begin{pmatrix} \langle \mathcal{T} \rangle & -\langle \chi \rangle \\ \chi(0) & 0 \end{pmatrix} , \quad \mathcal{T} := \mathcal{M}^{-1} H_{yy}^0 \mathcal{M}^{-T} , \quad \chi := \frac{1}{\rho} \mathcal{M}^{-1} H_y^0 , \quad H_\cdot^0 := H_\cdot(\theta + u, v) .$$

[4.11] The matrix $\hat{\mathcal{A}}_0$ is related to but different from the matrix $\mathcal{A}_0$ in (4.1.1).

By (4.2.4), the norm of $\mathcal{B} = \sum_{k\geq 1}(-\partial_\theta u^{(0)})^k$ can be bounded, for real θ, by[4.12]

$$\|\mathcal{B}\|_{0,\mathcal{E}} \leq \|u_\theta^{(0)}\|_{0,\mathcal{E}}\,(1-\|u_\theta^{(0)}\|_{0,\mathcal{E}})^{-1} \leq \tilde{U}^{(0)\pm}_{\mathrm{real}}\,\big(1-\tilde{U}^{(0)\pm}_{\mathrm{real}}\big)^{-1}$$

Recalling that $H(x,y) = H_0(y) + \varepsilon H_1(x,y)$ and (4.2.4), we see that the norm of $\mathcal{C}$ can be bounded, for real θ, by

$$\begin{aligned}\|\mathcal{C}\|_{0,\mathcal{E}} &\leq \|\partial_y^3 H_0\|_r\,\|v^{(0)}-y_0\|_{0,\mathcal{E}} + \varepsilon_0\|\partial_y^2 H_1\|_{\bar{\xi},r}\\ &\leq \|\partial_y^3 H_0\|_r\,\check{V}^{(0)\pm}_{\mathrm{real}} + \varepsilon_0\|\partial_y^2 H_1\|_{\bar{\xi},r}\ .\end{aligned}$$

The estimate on the norm of b, for real θ, is similar and one gets

$$\|b\|_{0,\mathcal{E}} \leq \|\partial_y^2 H_0\|_r\,\check{V}^{(0)\pm}_{\mathrm{real}} + \varepsilon_0\|\partial_y H_1\|_{\bar{\xi},r} + \|\mathcal{B}\|_{0,\mathcal{E}}\,E_{0,1}\ . \tag{4.2.10}$$

Next, we estimate

$$\begin{aligned}\|\mathcal{T}-\partial_y^2 H_0(y_0)\|_{0,\mathcal{E}} &= \|\partial_y^2 H_0(y_0)\mathcal{B}^T + \big(\mathcal{C}+\mathcal{B}\partial_y^2 H^0\big)\big(I+\mathcal{B}^T\big)\|_{0,\mathcal{E}}\\ &\leq \|\mathcal{B}\|_{0,\mathcal{E}}\,|\partial_y^2 H_0(y_0)| + \big(\|\mathcal{C}\|_{0,\mathcal{E}} + \|\mathcal{B}\|_{0,\mathcal{E}}\,E_{0,2}\big)\big(1+\|\mathcal{B}\|_{0,\mathcal{E}}\big)\ .\end{aligned} \tag{4.2.11}$$

Notice that

$$|\partial_y^2 H_0(y_0)| = \frac{3}{L_\pm^4}\ ,$$

while the estimate on $\|\partial_y^q H_0\|_r$ and $\|\partial_y^q H_1\|_{\bar{\xi},r}$ have been discussed in 4.2.2. Since (recall (2.5.3) and the definition of b in (4.2.9)),

$$|\mathcal{A}_1|_{\mathcal{E}} \leq \sqrt{\big|\langle\mathcal{T}\rangle - \partial_y^2 H_0(y_0)\big|^2 + 2\left(\frac{\|b\|_{0,\mathcal{E}}}{\rho}\right)^2}\ ,$$

(4.2.11), (4.2.10) allow to bound $\mathcal{A}_1$ in (4.2.6). A bound on $|\mathcal{A}^{-1}|_{\mathcal{E}}$ now follows from (4.2.7) and (4.2.8).

Implementing this discussion one obtains the following numerical values[4.13]*:*

$$\begin{aligned}
&\|\partial_y^2 H_0^+\|_r \leq 15.384979...\ , && \|\partial_y^3 H_0^+\|_{\bar{\xi}} \leq 119.267702...\ ,\\
&\|\partial_y^2 H_0^-\|_r \leq 15.335556...\ , && \|\partial_y^3 H_0^-\|_{\bar{\xi}} \leq 159.060236...\ ,\\
&\|\partial_y H_1^+\|_{\bar{\xi},r} \leq 1.794947...\cdot 10^{-2}\ , && \|\partial_y^2 H_1^+\|_{\bar{\xi},r} \leq 0.366425...\ ,\\
&\|\partial_y H_1^-\|_{\bar{\xi},r} \leq 1.938072...\cdot 10^{-2}\ , && \|\partial_y^2 H_1^-\|_{\bar{\xi},r} \leq 0.519666...\ ,\\
&\|\mathcal{B}^+\|_{0,\mathcal{E}} \leq 1.319218...\cdot 10^{-2}\ , && \|\mathcal{C}^+\|_{0,\mathcal{E}} \leq 0.403758...\ ,\\
&\|\mathcal{B}^-\|_{0,\mathcal{E}} \leq 1.450716...\cdot 10^{-2}\ , && \|\mathcal{C}^-\|_{0,\mathcal{E}} \leq 0.567616...\ ,\\
&\overline{A}^{(0)+} = 4.639466...\ , && \overline{A}^{(0)-} = 42.991575...\ .
\end{aligned}$$

[4.12] Recall (4.2.5).

[4.13] We reinsert, when necessary, the suffix $\pm$.

4.2.5. Evaluation of $F^{(0)\pm}$, $G^{(0)\pm}$, $\bar{h}^{(0)\pm}$.

By definition, $(u^{(0)}, v^{(0)}, \omega^{(0)}) = (u^{(0)\pm}, v^{(0)\pm}, \omega^{(0)\pm})$ solves, up to order 12 in ε, the system (2.2.6) with $\omega = \big(1 + a(\varepsilon)\big)\omega_\pm$ and $E = E_{\rm V}(\varepsilon) = E_{\rm V}^{(0)} + \varepsilon E_{\rm V}^{(1)}$. Thus, inserting $(u^{(0)}, v^{(0)}, \omega^{(0)})$ into (2.3.1), we find[4.14]:

$$
\begin{aligned}
\omega^{(0)} + D_{\omega^{(0)}} u^{(0)} - H_y(\theta + u^{(0)}, v^{(0)}) &= \sum_{k=13}^{24} \varepsilon^k \sum_{h=k-12}^{12} a_h D_{\omega_\pm} u_{k-h} \\
&\quad - \sum_{k=13}^{\infty} \Big[H_y(\theta + u^{(0)}, v^{(0)})\Big]_k \varepsilon^k \\
&=: \sum_{k=13}^{24} \varepsilon^k \sum_{h=k-12}^{12} a_h D_{\omega_\pm} u_{k-h} + \tilde{f}^{(0)}(\theta;\varepsilon) \\
&=: f^{(0)}(\theta;\varepsilon) \ ; \\
D_{\omega^{(0)}} v^{(0)} + H_x(\theta + u^{(0)}, v^{(0)}) &= \sum_{k=13}^{24} \varepsilon^k \sum_{h=k-12}^{12} a_h D_{\omega_\pm} v_{k-h} \\
&\quad + \sum_{k=13}^{\infty} \Big[H_x(\theta + u^{(0)}, v^{(0)})\Big]_k \varepsilon^k \\
&= \sum_{k=13}^{24} \varepsilon^k \sum_{h=k-12}^{12} a_h D_{\omega_\pm} v_{k-h} + \tilde{g}^{(0)}(\theta;\varepsilon) \\
&= g^{(0)}(\theta;\varepsilon) \ ; \\
u^{(0)}(0) &= 0 \ ; \\
H(0, v^{(0)}(0)) - \big(E_{\rm V}^{(0)} + \varepsilon E_{\rm V}^{(1)}\big) &= \sum_{k=13}^{\infty} \Big[H(0, v^{(0)}(0))\Big]_k \varepsilon^k \\
&=: h^{(0)} \ .
\end{aligned}
$$

We recall that $F^{(0)}$, $G^{(0)}$ and $\bar{h}^{(0)}$ are upper bounds on, respectively $\|f^{(0)}\|_{\xi^{(0)},\mathcal{E}}$, $\|g^{(0)}\|_{\xi^{(0)},\mathcal{E}}$ and $|h^{(0)}|_{\mathcal{E}}$. We begin by estimating

$$
\begin{aligned}
\|f^{(0)}\|_{\xi^{(0)},\mathcal{E}} &\le \sum_{k=13}^{24} \varepsilon_0^k \sum_{h=k-12}^{12} |a_h| \, \|D_{\omega_\pm} u_{k-h}\|_{\xi^{(0)}} + \|\tilde{f}^{(0)}\|_{\xi^{(0)},\mathcal{E}} \ ; \\
\|g^{(0)}\|_{\xi^{(0)},\mathcal{E}} &\le \sum_{k=13}^{24} \varepsilon_0^k \sum_{h=k-12}^{12} |a_h| \, \|D_{\omega_\pm} v_{k-h}\|_{\xi^{(0)}} + \|\tilde{g}^{(0)}\|_{\xi^{(0)},\mathcal{E}} \ .
\end{aligned}
$$

Similarly to what done in (4.2.3), an estimate of $\|D_{\omega_\pm} u_{k-h}\|_{\xi^{(0)}}$ and $\|D_{\omega_\pm} v_{k-h}\|_{\xi^{(0)}}$, follows from (4.1.8):

$$
\begin{aligned}
\|D_{\omega_\pm} u_{k-h}\|_{\xi^{(0)}} &\le \sum_{n \in \mathcal{N}_k^u} |\omega_\pm \cdot n| \, |u_n^{k\pm}| \exp(|n|\xi^{(0)}) \ ; \\
\|D_{\omega_\pm} v_{k-h}\|_{\xi^{(0)}} &\le \sum_{n \in \mathcal{N}_k^v} |\omega_\pm \cdot n| \, |v_n^{k\pm}| \exp(|n|\xi^{(0)}) \ .
\end{aligned}
$$

[4.14] Recall that $u^{(0)}$ is a sum of sines and therefore odd in θ.

We turn, now, to discuss estimates on $\|\tilde f^{(0)}\|_{\xi^{(0)},\mathcal{E}}$, $\|\tilde g^{(0)}\|_{\xi^{(0)},\mathcal{E}}$ and $|h^{(0)}|_{\mathcal{E}}$, which are based upon Lemma 2.5.2.

Recalling once more Remark 4.1.3, we see that the form of (the components of) $\tilde f^{(0)}$, $\tilde g^{(0)}$ and $h^{(0)}$ is equal to the form of (sum of) $\tilde f_M$ in (2.5.18) with: $j \le M = 12$, $a^{(j)} = u_j^{\pm}$, $b^{(j)} = v_j^{\pm}$, $f(x,y) = \sum_{n\in\mathcal{N}} f_n(y)c_n(x)$ being given by H_i, $\partial_{x_k}H_i$ or $\partial_{y_k}H_i$ for $i = 0,1$ and $k = 1,2$. For example

$$\tilde f_1^{(0)} = -\sum_{k\ge 13}\left[\frac{1}{\left(v_1^{(0)}\right)^3}\right]_k \varepsilon^k - \sum_{k\ge 13}\left[\partial_{y_1}H_1(\theta+u^{(0)},v^{(0)})\right]_{k-1}\varepsilon^k\,, \tag{4.2.12}$$

and the "f" (compare (2.5.21) and (2.5.18)) corresponding to the first sum in (4.2.12) is simply

$$f(x,y) = -\frac{1}{y_1^3}\,,$$

while the "f" corresponding to the second sum in (4.2.12) is given by

$$f(x,y) = \sum_{|n|_1\le 10} f_n(y)\cos(n\cdot x)\,,$$

with the explicit form of the f_n's deduced by differentiating with respect to y_1 the function $-H_1 = P_{\rm SJV}$ given in (3.4.5) (where $(L,G) = y$). In general, the form of the coefficients a_n has been spelled out in Remark 4.1.3: see, in particular, (4.1.5).

The functions F_n needed in order to apply Lemma 2.5.2 are obtained by majorizing the above mentioned f_n's and can be obtained by using (2.5.15) in § 2.5.6. For example,

$$\pm\frac{1}{y_1^3} \prec \left(\frac{1}{y_{01}-(y_1-y_{01})}\right)^3$$

or

$$c\,y_2^2\,e(y)^{-3} \prec |c|\,\big(y_{02}+(y_2-y_{02})\big)^2\,\tilde E(y)^3\,.$$

The numbers $A_n^{(j)}$ and $B_i^{(j)}$ are immediately computed using (4.1.8): for $n\in\mathbb{Z}^2$ we find

$$\begin{aligned}
\|n\cdot u_j^{\pm}\|_{\xi^{(0)}} &= \Big\|n\cdot\sum_{m\in\mathcal{N}_j^u} u_m^{j\pm}\sin(m\cdot\theta)\Big\|_{\xi^{(0)}}\\
&\le \sum_{m\in\mathcal{N}_j^u} |n\cdot u_m^{j\pm}|\exp(|m|\xi^{(0)})\\
&=: A_n^{(j)}\,;
\end{aligned}$$

and, for $i = 1,2$,

$$\begin{aligned}
\|v_{ji}^{\pm}\|_{\xi^{(0)}} &= \Big\|\sum_{m\in\mathcal{N}_j^v} v_{mi}^{j\pm}\cos(m\cdot\theta)\Big\|_{\xi^{(0)}}\\
&\le \sum_{m\in\mathcal{N}_j^v} |v_{mi}^{j\pm}|\exp(|m|\xi^{(0)})\\
&=: B_i^{(j)}\,.
\end{aligned} \tag{4.2.13}$$

At this point the procedure should be clear. As an explicit (simple) example, an estimate, based on Lemma 2.5.2, of the first sum in (4.2.12) is:

$$\Big\| - \sum_{k\geq 13} \Big[\frac{1}{\big(v_1^{(0)\pm}\big)^3}\Big]_k \varepsilon^k \Big\|_{\xi^{(0)},\mathcal{E}} \leq \Big(\frac{1}{L_\pm - \sum_{j=1}^{12} B_1^{(j)} \varepsilon_0^j}\Big)^3 - \sum_{j=0}^{12} \Big[\Big(\frac{1}{L_\pm - \sum_{j=1}^{12} B_1^{(j)} \varepsilon^j}\Big)^3\Big]_j \varepsilon_0^j$$

where the $B_1^{(j)}$ is as in (4.2.13). Observe that the explicit computation of the 13 numbers

$$\Big[\Big(\frac{1}{L_\pm - \sum_{j=1}^{12} B_1^{(j)} \varepsilon^j}\Big)^3\Big]_j \,, \qquad (0 \leq j \leq 12)$$

may be trivially implemented on a machine.

Implementing this discussion one obtains the following numerical values:

$$\begin{aligned} F^{(0)+} &\leq 7.463545... \cdot 10^{-22} \,, & F^{(0)-} &\leq 8.488723... \cdot 10^{-21} \,, \\ G^{(0)+} &\leq 6.279188... \cdot 10^{-18} \,, & G^{(0)-} &\leq 7.520731... \cdot 10^{-17} \,, \\ \bar{h}^{(0)+} &\leq 1.389888... \cdot 10^{-23} \,, & \bar{h}^{(0)-} &\leq 1.516977... \cdot 10^{-22} \,. \end{aligned}$$

The evaluation of the input parameters for the KAM norm map associated to the approximate tori $\big(u^{(0)\pm}, v^{(0)\pm}, \omega^{(0)\pm}\big)$ has been completed.

4.3. Iterations of the KAM map

In this section we iterate the KAM norm map $\widehat{\mathcal{K}}$ defined in Proposition 2.6.4; recall, also, Remark 2.6.5. In fact, we shall iterate $\mathcal{K}$ *two times* in order to construct the tori $\mathcal{T}_\varepsilon^+$, for $|\varepsilon| \leq 10^{-3}$, and *four times* in order to construct the tori $\mathcal{T}_\varepsilon^-$.

We recall that the KAM norm map $\widehat{\mathcal{K}}$ (besides depending upon the fixed parameters κ and ρ) depends explicitly on the "analyticity losses" δ_i *and* on computable upper bounds $\sigma_{p,k}$ on the "small divisors" $s_{1,0}$ and $s_{p,1}$ with $p = 0, 1, 2$; see (2.6.5) and footnote 2.16.

We begin by defining the ($\pm$-independent) "analyticity losses":

$$\begin{aligned} \delta_0 &:= 1.237500... \cdot 10^{-2} \,, & \delta_1 &:= 6.187500... \cdot 10^{-3} \,, \\ \delta_2 &:= 3.093750... \cdot 10^{-3} \,, & \delta_3 &:= 1.546875... \cdot 10^{-3} \,, \end{aligned}$$

so that, since

$$\xi^{(i+1)} := \xi^{(i)} - 2\delta_i \,, \qquad 0 \leq i \leq 3 \,,$$

we find

$$\begin{aligned} \xi^{(1)} &:= 2.575000... \cdot 10^{-2} \,, & \xi^{(2)} &:= 1.337500... \cdot 10^{-2} \,, \\ \xi^{(3)} &:= 7.187500... \cdot 10^{-3} \,, & \xi^{(4)} &:= 4.093750... \cdot 10^{-3} \,. \end{aligned}$$

The $\sigma_{p,k}$ shall be taken as follows. First, observe that, since

$$\omega^{(0)\pm} = (1 + a_\pm)\, \omega_\pm$$

with $(1+a_\pm)>0$ (compare (4.1.9), (4.1.10) and (4.1.11)), from the definition of $s_{p,k}(\delta,\omega)$, (2.5.23), one has that

$$s_{p,k}(\delta;\omega^{(0)\pm})=(1+a_\pm)\,s_{p,k}(\delta;\omega_\pm)\ .$$

Therefore, we turn to the definition of $\sigma_{p,k}(\delta,\omega_\pm)$. In view of the form of $\omega_\pm$ (see (3.4.8), (3.4.7)), by Proposition B.1.1 (see, in particular, (B.1.2)) , we have that

$$|\omega_+\cdot n|\geq\frac{\gamma_{\rm g}}{|n_1|}\ ,\qquad\qquad \forall|n_1|\geq N_+\ ,$$
$$|\omega_-\cdot n|\geq\frac{\gamma_{\rm g}}{|n_1|}\ ,\qquad\qquad \forall|n_1|\geq N_-\ ,$$

$\gamma_{\rm g}=2/(3+\sqrt{5})$ and

$$N_+=321\ ,\qquad\qquad N^-=71\ .$$

Thus, by Proposition B.2.1, we can take

$$\begin{aligned}\sigma_{0,1}(\delta;\omega_\pm)\ &:=\ \frac{\sqrt{2}}{e\,\delta}\ ,\\ \sigma_{p,0}(\delta;\omega_\pm)\ &:=\ \max\Big\{1\ ,\ \Big(\frac{p}{E}\Big)^p\ \frac{1}{(\gamma_{\rm g})^p}\ \frac{1}{\delta^p}\ ,\ \max_{\substack{1\leq n_1<N^\pm\\ 0\leq n_2<\alpha_\pm N^\pm+1}}\frac{\exp(-\delta|n|)}{|\alpha_\pm n_1-n_2|^p}\Big\}\ ,\\ \sigma_{p,1}(\delta;\omega_\pm)\ &:=\ \sqrt{\hat\sigma_{p,e_1}(\delta)^2+\hat\sigma_{p,e_2}(\delta)^2}\ ,\end{aligned}$$

where, denoting $\mathcal{N}_{N^\pm}:=\{n:\ 1\leq n_1<N^\pm \text{ and } 0\leq n_2<\alpha_{\pm N^\pm+1}\}$, it is:

$$\hat\sigma_{p,e_1}(\delta;\omega_\pm):=\max\Big\{\frac{1}{e\,\delta}\ ,\ \Big(\frac{p+1}{E}\Big)^{p+1}\ \frac{1}{(\gamma_{\rm g})^p}\ \frac{1}{\delta^{p+1}}\ ,\ \max_{\mathcal{N}_{N^\pm}}\frac{n_1\ \exp(-\delta|n|)}{|\alpha_\pm n_1-n_2|^p}\Big\}\ ,$$
$$\hat\sigma_{p,e_2}(\delta;\omega_\pm):=\max\Big\{\frac{1}{e\,\delta}\ ,\ \frac{2^{\frac{p+1}{2}}}{E}\Big(\frac{p}{E}\Big)^p\ \frac{1}{(\gamma_{\rm g})^p}\ \frac{1}{\delta^{p+1}}\ ,\ \max_{\mathcal{N}_{N^\pm}}\frac{n_2\ \exp(-\delta|n|)}{|\alpha_\pm n_1-n_2|^p}\Big\}\ .$$

Recall that by construction (see (2.4.4))

$$\omega^{(i)\pm}=\Big(\prod_{j=0}^{i-1}(1+a^{(j)\pm})\Big)(1+a_\pm)\,\omega_\pm\ ; \tag{4.3.1}$$

recall also that (compare (2.6.17) and (2.6.16))

$$|a^{(j)\pm}|\leq\frac{\eta_{13}^{(j)\pm}}{\rho}\ . \tag{4.3.2}$$

We, therefore, see that, as explicit upper bounds

$$\sigma_{p,k}^{(i)\pm}(\delta):=\sigma_{p,k}\big(\delta;\omega^{(i)\pm}\big)$$

for the $i^{\rm th}$ iteration, one can recursively choose the following numbers:

$$\begin{aligned}\sigma_{p,k}^{(0)\pm}\ &:=(1+a_\pm)\,\sigma_{p,k}(\delta;\omega_\pm)\ ,\\ \sigma_{p,k}^{(i)\pm}\ &:=\left(\prod_{j=0}^{i-1}\Big(1+\frac{\eta_{13}^{(j)\pm}}{\rho}\Big)\right)\,(1+a_\pm)\,\sigma_{p,k}(\delta;\omega_\pm)\ .\end{aligned}$$

We, then, set for $i\geq 0$,

$$\mathcal{K}\big(u^{(i)},v^{(i)},\omega^{(i)}\big)=:\big(u^{(i+1)},v^{(i+1)},\omega^{(i+1)}\big)\ , \tag{4.3.3}$$

and

$$\widehat{\mathcal{K}}_{\delta_i,\sigma^{(i)}_{p,k}}\big(\xi^{(i)},\gamma^{(i)},F^{(i)},G^{(i)},\bar{h}^{(i)},M^{(i)},\overline{M}^{(i)},U^{(i)},V^{(i)},\tilde{V}^{(i)},\overline{A}^{(i)}\big)$$
$$=: \big(\xi^{(i+1)},\gamma^{(i+1)},F^{(i+1)},G^{(i+1)},\bar{h}^{(i+1)},M^{(i+1)},\overline{M}^{(i+1)},U^{(i+1)},$$
$$V^{(i+1)},\tilde{V}^{(i+1)},\overline{A}^{(i+1)}\big)\ ,$$

provided (recall point (ii) of Remark 2.6.5)

$$\begin{aligned}
&\overline{M}^{(i)}\Big(\eta_2^{(i)}+\rho\eta_0^{(i)}+\frac{\eta_0^{(i)}\eta_5^{(i)}}{1-\eta_0^{(i)}}\Big)+\sum_{j=0}^{i-1}\eta_{25}^{(j)}\le r-|v^{(0)}(0)-y_0|_{\infty,\varepsilon}\ ,\\
&(\eta_{10}^{(i)})^2\le(\kappa-1)\eta_8^{(i)}\ ,\\
&B^{(i)}\eta_{11}^{(i)}<1\ ,\\
&\eta_{13}^{(i)}<\rho\ ,\\
&\overline{M}^{(i)}\eta_{24}^{(i)}<1\ ,\\
&\overline{A}^{(i)}\sqrt{(\eta_{30}^{(i)})^2+(\eta_{31}^{(i)})^2+\eta_{30}^{(i)}\eta_{31}^{(i)}}<1\ ,\\
&\eta_{23}^{(i)}<2\delta_i\ ,\\
&\sum_{j=0}^{i}\eta_{25}^{(j)}<r-\sup_{|\varepsilon|\le\varepsilon_0}\ \sup_{\theta\in\mathbb{T}^2_{\xi^{(0)}}}|v^{(0)}(\theta)-y_0|_\infty
\end{aligned}\tag{4.3.4}$$

are satisfied. We will see, below, that (4.3.4) are recursively satisfied for $0\le i\le 1$ in the "+ case" and for $0\le i\le 3$ on the "− case".

In fact, implementing this discussion one obtains the following numerical values:

$$|v^{(0)+}(0)-y^+|_\varepsilon\le\sum_{k=1}^{12}\Big|\sum_{n\in\mathcal{N}_k^v}v_n^{k+}\Big|\varepsilon_0^k\le 5.051313...\cdot10^{-5}\ ,$$
$$\|v^{(0)+}-y^+\|_{0,\varepsilon}\le\sum_{k=1}^{12}\sum_{n\in\mathcal{N}_k^v}|v_n^{k+}|\varepsilon_0^k\le 3.529186...\cdot10^{-4}\ ,$$

$$|v^{(0)-}(0)-y^-|_\varepsilon\le\sum_{k=1}^{12}\Big|\sum_{n\in\mathcal{N}_k^v}v_n^{k-}\Big|\varepsilon_0^k\le 6.291231...\cdot10^{-5}\ ,$$
$$\|v^{(0)-}-y^-\|_{0,\varepsilon}\le\sum_{k=1}^{12}\sum_{n\in\mathcal{N}_k^v}|v_n^{k-}|\varepsilon_0^k\le 3.382850...\cdot10^{-4}\ ,$$

$$\begin{aligned}
&\gamma^{(1)+}\ge 3.324329...\cdot10^{-2}\ , && F^{(1)+}\le 2.411675...\cdot10^{-23}\ ,\\
&G^{(1)+}\le 1.264526...\cdot10^{-20}\ , && \bar{h}^{(1)+}=0\ ,\\
&M^{(1)+}\le 1.015797...\ , && \overline{M}^{(1)+}\le 1.016051\ ,\\
&U^{(1)+}\le 3.772624...\cdot10^{-3}\ , && V^{(1)+}\le 0.933497...\ ,\\
&\tilde{V}^{(1)+}\le 1.292660...\cdot10^{-3}\ , && \overline{A}^{(1)+}\le 4.639476...\ ;
\end{aligned}$$

$$\begin{aligned}
&\gamma^{(1)-} \geq 7.224496... \cdot 10^{-3} , && F^{(1)-} \leq 9.355917... \cdot 10^{-20} , \\
&G^{(1)-} \leq 5.034106... \cdot 10^{-17} , && \bar{h}^{(1)-} = 0 , \\
&M^{(1)-} \leq 1.017377... , && \overline{M}^{(1)-} \leq 1.017684... , \\
&U^{(1)-} \leq 4.149878... \cdot 10^{-3} , && V^{(1)-} \leq 0.939554... , \\
&\tilde{V}^{(1)-} \leq 1.256448... \cdot 10^{-3} , && \overline{A}^{(1)-} \leq 43.044043 ;
\end{aligned}$$

$$\begin{aligned}
&\gamma^{(2)-} \geq 7.224496... \cdot 10^{-3} , && F^{(2)-} \leq 1.524165... \cdot 10^{-23} , \\
&G^{(2)-} \leq 9.690154... \cdot 10^{-21} , && \bar{h}^{(2)-} = 0 , \\
&M^{(2)-} \leq 1.017377... , && \overline{M}^{(2)-} \leq 1.017684... , \\
&U^{(2)-} \leq 4.149878... \cdot 10^{-3} , && V^{(2)-} \leq 0.939554... , \\
&\tilde{V}^{(2)-} \leq 1.256448... \cdot 10^{-3} , && \overline{A}^{(2)-} \leq 43.045127... ;
\end{aligned}$$

$$\begin{aligned}
&\gamma^{(3)-} \geq 7.224496... \cdot 10^{-3} , && F^{(3)-} \leq 9.585046... \cdot 10^{-30} , \\
&G^{(3)-} \leq 6.720494... \cdot 10^{-27} , && \bar{h}^{(3)-} = 0 , \\
&M^{(3)-} \leq 1.017377... , && \overline{M}^{(3)-} \leq 1.017684... , \\
&U^{(3)-} \leq 4.149878... \cdot 10^{-3} , && V^{(3)-} \leq 0.939554... , \\
&\tilde{V}^{(3)-} \leq 1.256448... \cdot 10^{-3} , && \overline{A}^{(3)-} \leq 43.045128... .
\end{aligned}$$

With such values one verifies immediately that (4.3.4) are satisfied for $0 \leq i \leq 1$ in the "+ case" and for $0 \leq i \leq 3$ in the "− case". Therefore one can compute also:

$$\begin{aligned}
&\gamma^{(2)+} \geq 3.324329... \cdot 10^{-2} , && F^{(2)+} \leq 1.386735... \cdot 10^{-32} , \\
&G^{(2)+} \leq 8.650985... \cdot 10^{-30} , && \bar{h}^{(2)+} = 0 , \\
&M^{(2)+} \leq 1.015797... , && \overline{M}^{(2)+} \leq 1.016051... , \\
&U^{(2)+} \leq 3.772624... \cdot 10^{-3} , && V^{(2)+} \leq 0.933497... , \\
&\tilde{V}^{(2)+} \leq 1.292660... \cdot 10^{-3} , && \overline{A}^{(2)+} \leq 4.639476... ;
\end{aligned}$$

$$\begin{aligned}
&\gamma^{(4)-} \geq 7.224496... \cdot 10^{-3} , && F^{(4)-} \leq 9.194472... \cdot 10^{-41} , \\
&G^{(4)-} \leq 6.801231... \cdot 10^{-38} , && \bar{h}^{(4)-} = 0 , \\
&M^{(4)-} \leq 1.017377... , && \overline{M}^{(4)-} \leq 1.017684... , \\
&U^{(4)-} \leq 4.149878... \cdot 10^{-3} , && V^{(4)-} \leq 0.939554... , \\
&\tilde{V}^{(4)-} \leq 1.256448... \cdot 10^{-3} , && \overline{A}^{(4)-} \leq 43.045128... .
\end{aligned}$$

These parameters will be used in the next step as input parameters for Theorem 2.7.1 for, respectively, the "+ case" and the "− case".

For later use, we report also the estimates measuring the distance between, respectively, $\left(u^{(2)+}, v^{(2)+}\right)$, $\left(u^{(4)-}, v^{(4)-}\right)$ from the Lindstedt polynomials $\left(u^{(0)+}, v^{(0)+}\right)$, $\left(u^{(0)-}, v^{(0)-}\right)$ defined in (4.1.9). Recalling the definition of the KAM map (2.4.4) one has that

$$u^{(k)\pm} = u^{(0)\pm} + \sum_{i=0}^{k-1} z_i^{\pm} , \qquad v^{(k)\pm} = v^{(0)\pm} + \sum_{i=0}^{k-1} w_i^{\pm} ,$$

and recalling that from that the norm of z_i is bounded by $\eta_{23}^{(i)}$ and that the norm of w_i is bounded by $\eta_{25}^{(i)}$ (compare (2.6.19) and (2.6.20)), one obtains the following bounds:

$$\|u^{(2)+} - u^{(0)+}\|_{\xi^{(2)}} \le \sum_{i=0}^{1} \eta_{23}^{(i)+} \le 8.802982 \cdot 10^{-11} ,$$

$$\|v^{(2)+} - v^{(0)+}\|_{\xi^{(2)}} \le \sum_{i=0}^{1} \eta_{25}^{(i)+} \le 1.388253 \cdot 10^{-13} , \tag{4.3.5}$$

for the "+ case" and, for the "− case":

$$\|u^{(4)-} - u^{(0)-}\|_{\xi^{(4)}} \le \sum_{i=0}^{3} \eta_{23}^{(i)-} \le 5.332983 \cdot 10^{-9} ,$$

$$\|v^{(4)-} - v^{(0)-}\|_{\xi^{(4)}} \le \sum_{i=0}^{3} \eta_{25}^{(i)-} \le 8.353437 \cdot 10^{-12} . \tag{4.3.6}$$

4.4. Application of the iso-energetic KAM theorem and perpetual stability of the Sun-Jupiter-Victoria problem

We are, now, in a position to apply the iso-energetic KAM theorem 2.7.1 to the Sun-Jupiter-Victoria problem. We shall treat simultaneously the "+ case" and the "− case", i.e., the construction of the tori $\mathcal{T}_\varepsilon^+$ and $\mathcal{T}_\varepsilon^-$.

As approximate solution (u, v, ω) in Theorem 2.7.1, we take

$$(u, v, \omega) := \left(u^{(2)+}, v^{(2)+}, \omega^{(2)+}\right) , \qquad \left(\text{“} + \text{case”}\right) ,$$

$$(u, v, \omega) := \left(u^{(4)-}, v^{(4)-}, \omega^{(4)-}\right) , \qquad \left(\text{“} - \text{case”}\right) , \tag{4.4.1}$$

defined in (4.3.3). Then, (f, g, h) is defined as, respectively, $\left(f^{(2)+}, g^{(2)+}, 0\right)$ and $\left(f^{(4)-}, g^{(4)-}, 0\right)$. The other parameters are, consistently, taken to be as follows: $\bar{\xi}^{\pm}$, $r^{\pm}$, κ, ρ, ε_0 are as in § 4.2.1; $y_0 = y^{\pm}$, $E := E_0$ as in (4.1.2); $E_{p,q}$ are as in § 4.2.2;

$$\gamma = \gamma^{(k)\pm} ; \qquad \xi = \xi^{(k)\pm} ; \qquad F = F^{(k)\pm} ; \qquad G = G^{(k)\pm} ; \qquad h = \bar{h} = 0 ;$$
$$M = M^{(k)\pm} ; \qquad \overline{M} = \overline{M}^{(k)\pm} ; \qquad \overline{A} = \overline{A}^{(k)\pm} ; \qquad \tilde{V} = \tilde{V}^{(k)\pm} ;$$

with $k = 2$ in the "+ case" and $k = 4$ in the "− case". Recalling (4.3.1) and (4.3.2), we see that as parameter Ω we can take

$$\Omega := \left(\prod_{i=0}^{k-1} \left(1 + \frac{\eta_{13}^{(i)\pm}}{\rho}\right) \right) (1 + a_\pm) \, |\omega_\pm| ,$$

with $k = 1$ in the "+ case" and $k = 3$ in the "− case". Now, observe that, since (4.3.4) are verified for, respectively, $i = 1$ and $i = 3$, by Proposition 2.6.4, (2.6.33) and (2.6.34) hold, i.e.,

$$\sup_{\mathbb{T}^2_{\xi^{(k)\pm}}} \left|\operatorname{Im} u^{(k)\pm}\right| < \bar{\xi} - \xi^{(k)\pm} , \qquad \hat{r}^{\pm} := \sup_{\mathbb{T}^d_{\xi^{(k)\pm}}} |v^{(k)\pm}(\theta) - y_0|_\infty < r ,$$

for $k = 2$ and $k = 4$ in, respectively, the "+ case" and the "− case". Thus (2.7.3) in Theorem 2.7.1 is verified by our approximate tori (u, v, ω) in (4.4.1).

Next, we evaluate the weighted norms defined in (2.7.6). *Implementing this discussion one obtains the following numerical values:*

$$\begin{aligned}
&E_1^{*+} = 3.515464\ldots\,, \qquad E_2^{*+} = 15.384979\ldots\,,\\
&E_3^{*+} = 119.267702\ldots\,, \qquad E^{*+} = 4.770708\ldots\,,\\
&\Omega^{*+} = 4.770708\,,\\
&\beta_0^+ = 1\,, \qquad \beta_1^+ = 23.853540\ldots\,,\\
&\alpha^+ = 110.667935\ldots\,, \qquad \mu^+ = 2.482139\ldots \cdot 10^{-30}\,,
\end{aligned}$$

$$\begin{aligned}
&E_1^{*-} = 3.488445\ldots\,, \qquad E_2^{*-} = 15.335556\ldots\,,\\
&E_3^{*-} = 159.060236\ldots\,, \qquad E^{*-} = 6.362409\ldots\,,\\
&\Omega^{*-} = 6.362409\ldots\,,\\
&\beta_0^- = 1\,, \qquad \beta_1^- = 31.812047\ldots\,,\\
&\alpha^- = 1369.353670\,, \qquad \mu^- = 1.967516\ldots \cdot 10^{-38}\,.
\end{aligned}$$

Now, we fix ξ_∞ by letting

$$\xi_\infty := 10^{-3}\,, \tag{4.4.2}$$

so that (see (2.7.7))

$$\xi_*^+ = 10^{-3}\,, \qquad \xi_*^- = 7.734375\ldots \cdot 10^{-4}\,.$$

The constants c_*, c_{**} and $\hat{c}$, according to Lemma 2.7.4, can be taken to be

$$c_* = 38528.281271\ldots\,, \qquad c_{**} = 49.087867\ldots\,, \qquad \hat{c} = 111.699232\ldots\,.$$

At this point we can check the main hypotheses of Theorem 2.7.1, namely (2.7.8). *Implementing this discussion one obtains the following numerical values:*

$$\begin{aligned}
&[c_*\,\big(\overline{M}^{(2)+}\big)^{10}\big(M^{(2)+}\big)^4\,(\xi_*^+)^{-(4\tau+1)}\,\frac{\Omega^{*+}}{\Omega^+}(\alpha^+)^2(\beta_0^+)^4(\beta_1^+)^4]\;\mu^+ < \kappa_1,\\
&[c_{**}\,\big(\overline{M}^{(2)+}\big)^{5}\big(M^{(2)+}\big)^2\,(\xi_*^+)^{-2\tau}\,\alpha^+(\beta_0^+)^2(\beta_1^+)^2\,\frac{\rho}{r^+ - \hat{r}^+}]\;\mu^+ < \kappa_2\,;\\
&[c_*\,\big(\overline{M}^{(4)-}\big)^{10}\big(M^{(4)-}\big)^4\,(\xi_*^-)^{-(4\tau+1)}\,\frac{\Omega^{*-}}{\Omega^-}(\alpha^-)^2(\beta_0^-)^4(\beta_1^-)^4]\;\mu^- < \kappa_3\,,\\
&[c_{**}\,\big(\overline{M}^{(4)-}\big)^{5}\big(M^{(4)-}\big)^2\,(\xi_*^-)^{-2\tau}\,\alpha^-(\beta_0^-)^2(\beta_1^-)^2\,\frac{\rho}{r^- - \hat{r}^-}]\;\mu^- < \kappa_4\,,
\end{aligned}$$

where

$$\begin{aligned}
&\kappa_1 := 0.648020\,, \qquad \kappa_2 := 1.800024 \cdot 10^{-14}\,,\\
&\kappa_3 := 1.235897 \cdot 10^{-5}\,, \qquad \kappa_4 := 5.485636 \cdot 10^{-21}\,.
\end{aligned}$$

Thus, by Theorem 2.7.1 and its parameter-dependent case discussed in § 2.8.1, we obtain the following existence result.

THEOREM 4.4.1. *Let* $H = H_0 + \varepsilon H_1$, $E = E^{(0)} + \varepsilon E^{(1)}$; *let, respectively,* $(u, v, \omega) := \big(u^{(2)+}, v^{(2)+}, \omega^{(2)+}\big)$, $(u, v, \omega) := \big(u^{(4)-}, v^{(4)-}, \omega^{(4)-}\big)$; *let* ξ_∞ *be as above*[4.15]. *Then, for any* $|\varepsilon| \le 10^{-3}$, *there exists constants* $\tilde a^{\pm} \in \mathbb{R}$ *and real-analytic solutions* $(\tilde u^{\pm}, \tilde v^{\pm})$ *of the system*

$$\begin{aligned} \omega_{\tilde a^{\pm}} + D_{\omega_{\tilde a^{\pm}}} \tilde u^{\pm} - H_y(\theta + \tilde u^{\pm}, \tilde v^{\pm}) &= 0 \,, \\ D_{\omega_{\tilde a^{\pm}}} \tilde v^{\pm} + H_x(\theta + \tilde u^{\pm}, \tilde v^{\pm}) &= 0 \,, \\ \tilde u(0) &= 0 \,, \\ H(0, \tilde v^{\pm}(0)) &= E \,, \end{aligned}$$

where

$$\omega_{\tilde a^+} := (1 + \tilde a^+)\omega^{(2)+} = (1 + \tilde a^+)\Big(\prod_{i=0}^{1}(1 + a^{(i)+})\Big)\omega^+ \,,$$

$$\omega_{\tilde a^-} := (1 + \tilde a^-)\omega^{(4)-} = (1 + \tilde a^-)\Big(\prod_{i=0}^{3}(1 + a^{(i)-})\Big)\omega^- \,,$$

and

$$\max\Big\{|\tilde a^+| \,,\ \|\tilde u^+ - u^{(2)+}\|_{\xi_\infty} \,,\ \|\tilde u^+_\theta - u^{(2)+}_\theta\|_{\xi_\infty} \,,\ \frac{\|\tilde v^+ - v^{(2)+}\|_{\xi_\infty}}{\rho} \,,\ \frac{\|\tilde v^+_\theta - v^{(2)+}_\theta\|_{\xi_\infty}}{\rho}\Big\} \le 1.950724 \cdot 10^{-14} \,, \tag{4.4.3}$$

$$\max\Big\{|\tilde a^-| \,,\ \|\tilde u^- - u^{(4)-}\|_{\xi_\infty} \,,\ \|\tilde u^-_\theta - u^{(4)-}_\theta\|_{\xi_\infty} \,,\ \frac{\|\tilde v^- - v^{(4)-}\|_{\xi_\infty}}{\rho} \,,\ \frac{\|\tilde v^-_\theta - v^{(4)-}_\theta\|_{\xi_\infty}}{\rho}\Big\} \le 7.437412 \cdot 10^{-21} \,. \tag{4.4.4}$$

Furthermore $\tilde a^{\pm} = \tilde a^{\pm}(\varepsilon)$ *is real-analytic on the complex disc* $\{\varepsilon \in \mathbb{C} : |\varepsilon| \le 10^{-3}\}$ *and*

$$(\theta, \varepsilon) \to \big(\tilde u^{\pm}(\theta; \varepsilon), \tilde v^{\pm}(\theta; \varepsilon)\big)$$

is real-analytic on $\mathbb{T}^2_{\xi_\infty} \times \{\varepsilon \in \mathbb{C} : |\varepsilon| \le 10^{-3}\}$.
Finally, the Lindstedt polynomials $\big(u^{(0)+}, v^{(0)+}\big)$, $\big(u^{(0)-}, v^{(0)-}\big)$ *defined in* (4.1.9) *satisfy the bounds*

$$\begin{aligned} \|\tilde u^+ - u^{(0)+}\|_{\xi_\infty} &\le 8.804933 \cdot 10^{-11} \,, \\ \|\tilde v^+ - v^{(0)+}\|_{\xi_\infty} &\le 1.427268 \cdot 10^{-13} \,, \\ \|\tilde u^- - u^{(0)-}\|_{\xi_\infty} &\le 5.332984 \cdot 10^{-9} \,, \\ \|\tilde v^- - v^{(0)-}\|_{\xi_\infty} &\le 8.353438 \cdot 10^{-12} \,. \end{aligned} \tag{4.4.5}$$

The bounds (4.4.3) and (4.4.4) come from (2.7.9) after having estimated

$$\Big(\hat c\, \big(\overline{M}^{(3)+}\big)^5 \big(M^{(3)+}\big)^2 \xi_*^{-(2\tau+1)}\ \alpha^+ (\beta_0^+)^2 (\beta_1^+)^2\Big)\ \mu^+\ \le 1.950723... \cdot 10^{-14} \,,$$

$$\Big(\hat c\, \big(\overline{M}^{(4)-}\big)^5 \big(M^{(4)-}\big)^2 \xi_*^{-(2\tau+1)}\ \alpha^- (\beta_0^-)^2 (\beta_1^-)^2\Big)\ \mu^-\ \le 7.437411... \cdot 10^{-21} \,.$$

The statement about ε-analyticity follows from § 2.8.1. Estimates (4.4.5) come from (4.3.5), (4.3.6) and (4.4.3), (4.4.4). □

[4.15]Compare (4.1.1), (4.1.2), (4.3.3), (4.1.9)-(4.1.10) and (4.4.2).

To complete the proof of Claim 3.4.3, it remains to show that

$$(\ell(t), g(t), L(t), G(t)) := \phi^t_{H_{\rm SJV}|_{\varepsilon=\varepsilon_{\rm SJ}}}(\ell_0, g_0, L_{\rm V}, G_{\rm V})$$

belongs for any $t \in \mathbb{R}$ and any $(\ell_0, g_0) \in \mathbb{T}^2$ to the region $\mathcal{J}$. To prove this, we shall show that

$$\sup_{\mathbb{T}^2} \tilde{v}_1^+(\theta) < L_{\rm V} < \inf_{\mathbb{T}^2} \tilde{v}_1^-(\theta) \ , \qquad \sup_{\mathbb{T}^2} \tilde{v}_2^+(\theta) < G_{\rm V} < \inf_{\mathbb{T}^2} \tilde{v}_2^-(\theta) \ ,$$

which clearly implies the above statement. Recalling that $|w^{(i)}| \le \eta_{25}^{(i)}$, by Theorem 4.4.1 and the above definitions and estimates, we find:

$$\begin{aligned}\sup_{\theta\in\mathbb{T}^2} \tilde{v}_1^+(\theta) &\le L_+ + \Big(\sum_{k=1}^{12} 10^{-3k} \sum_{n\in\mathcal{N}_k^v} |v_{n1}^{k+}|\Big) + \sum_{i=1}^{2} \eta_{25}^{(i)+} + \rho\Big(\text{r.h.s. of (4.4.3)}\Big) \\ &< 0.669270 \ ;\end{aligned}$$

$$\begin{aligned}\sup_{\theta\in\mathbb{T}^2} \tilde{v}_2^+(\theta) &\le G_+ + \Big(\sum_{k=1}^{12} 10^{-3k} \sum_{n\in\mathcal{N}_k^v} |v_{n2}^{k+}|\Big) + \sum_{i=1}^{2} \eta_{25}^{(i)+} + \rho\Big(\text{r.h.s. of (4.4.3)}\Big) \\ &< 0.650639 \ ;\end{aligned}$$

and

$$\begin{aligned}\inf_{\theta\in\mathbb{T}^2} \tilde{v}_1^-(\theta) &\ge L_- - \Big[\Big(\sum_{k=1}^{12} 10^{-3k} \sum_{n\in\mathcal{N}_k^v} |v_{n1}^{k-}|\Big) + \sum_{i=1}^{4} \eta_{25}^{(i)-} + \rho\Big(\text{r.h.s. of (4.4.4)}\Big)\Big] \\ &> 0.670771 \ ,\end{aligned}$$

$$\begin{aligned}\inf_{\theta\in\mathbb{T}^2} \tilde{v}_2^-(\theta) &\ge G_- - \Big[[\Big(\sum_{k=1}^{12} 10^{-3k} \sum_{n\in\mathcal{N}_k^v} |v_{n2}^{k-}|\Big) + \sum_{i=1}^{4} \eta_{25}^{(i)-} + \rho\Big(\text{r.h.s. of (4.4.4)}\Big)\Big] \\ &> 0.656566 \ .\end{aligned}$$

We therefore see that

$$0.669270 < L_{\rm V} = 0.670 < 0.670771 \ , \qquad 0.650639 < G_{\rm V} = 0.654 < 0.656566 \ .$$

The proof of Claim 3.4.3 is completed. □

APPENDIX A

The Ellipse

In this appendix we recall a few classical facts about ellipses.

Cartesian equation. An ellipse is a *set of points in a plane with constant sum of distances from two given points*, called *foci*. The Cartesian equation of an ellipse, with respect to a reference plane $(x_1, x_2) \in \mathbb{R}^2$ with origin chosen as the middle point of the segment joining the two foci, is given by

$$\left(\frac{x_1}{a}\right)^2 + \left(\frac{x_2}{b}\right)^2 = 1 \ ,$$

where $2a$ is the (constant) sum of distances between x and the foci and

$$\left(\pm a\sqrt{1 - \left(\frac{b}{a}\right)^2}, 0\right) \tag{A.0.1}$$

are the coordinates of the foci. The positive numbers a and b are called, respectively, the *major* and the *minor semi-axis* of the ellipse; the number

$$e = \sqrt{1 - \left(\frac{b}{a}\right)^2}$$

is called the eccentricity of the ellipse. As it follows from (A.0.1), the distance c between one focus and the center $x = 0$ of the ellipse is given by

$$c = ea \ .$$

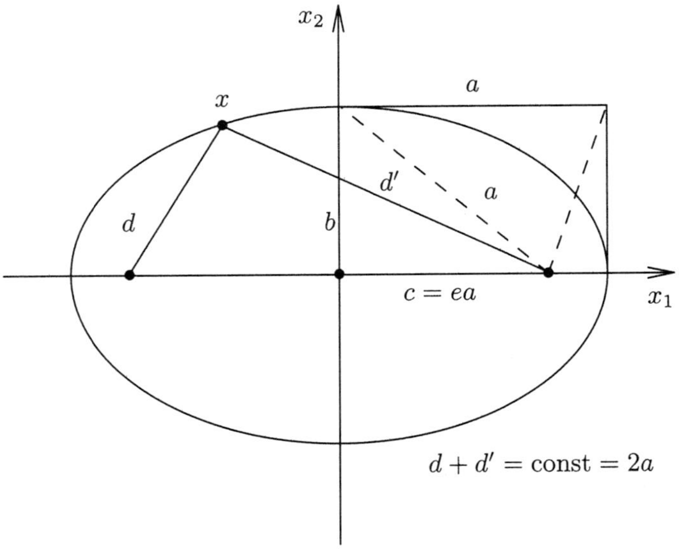

FIGURE 7. Ellipse of eccentricity 0.78.

Focal equation. Introducing polar coordinates (f, r) in the above x-plane taking as pole the focus $O = (c, 0)$, as f the angle between the x_1-axis and the axis joining O with the point x on the ellipse and $r = r(\mathrm{f})$ as the distance $|x - O|$ one finds the following *focal equation*

$$r = r(\mathrm{f}) := \frac{p}{1 + e\cos\mathrm{f}} ,$$

where p is called the *parameter* of the ellipse and is given by

$$p = a(1 - e^2) = \frac{b^2}{a} .$$

The angle f is called the *true anomaly.*

Parametric representation. The above ellipse is also described by the following parametric equations

$$x_1 = a\cos u , \qquad x_2 = b\sin u .$$

The angle u is called the *eccentric anomaly.*

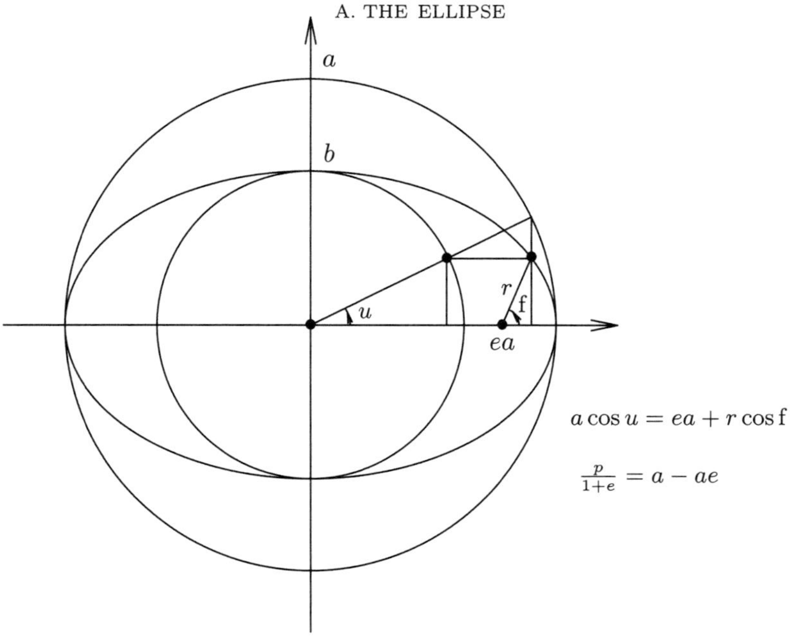

FIGURE 8. Ellipse parameters.

Thus a point x on the ellipse has the double representation:

$$x = (a\cos u, b\sin u) = (ea + r\cos \mathrm{f}, r\sin \mathrm{f}) \ ,$$

which relates the true and the eccentric anomalies. In particular, one finds:

$$\begin{aligned} & r\cos \mathrm{f} = a(\cos u - e) \ , \\ & r\sin \mathrm{f} = b\sin u = a\sqrt{1-e^2}\sin u \ , \\ & r = a(1 - e\cos u) \ , \\ & \tan\frac{\mathrm{f}}{2} = \sqrt{\frac{1+e}{1-e}}\tan\frac{u}{2} \ , \end{aligned}$$

$$\text{(A.0.2)} \qquad \mathrm{Area}(\mathcal{E}(\mathrm{f})) = \frac{ab}{2}(u - e\sin u) \ ,$$

where

$$\mathcal{E}(\mathrm{f}) := \{x = x(r', \mathrm{f}') : 0 \le r' \le r(\mathrm{f}) \ , 0 \le \mathrm{f}' \le \mathrm{f}\} \ .$$

APPENDIX B

Diophantine Estimates

B.1. Diophantine estimates for special quadratic numbers

Consider the vector

$$\omega := (\alpha, 1) \in \mathbb{R}^2 ,$$

where α is an irrational positive number having a continued fraction expansion[2.1] of the form

$$\begin{aligned} \alpha &= [a_0; a_1, ..., a_{j_0}, 1, 1, 1, ...] =: [a_0; a_1, ..., a_{j_0}, 1^\infty] \\ &= a_0 + \cfrac{1}{a_1 + \cfrac{1}{\ddots + \cfrac{1}{a_{j_0} + \cfrac{1}{1 + \ddots}}}} , \end{aligned} \tag{B.1.1}$$

for some $j_0 \in \mathbb{N}$ and $a_i \in \mathbb{N}$; this means that $a_j \equiv 1$ for $j \geq j_0 + 1$. If $a_0 = 0 = j_0$ then α is the so-called *golden mean*

$$\alpha_{\mathrm{g}} = \cfrac{1}{1 + \cfrac{1}{\ddots + \cfrac{1}{1 + \cfrac{1}{1 + \ddots}}}} = \frac{\sqrt{5} - 1}{2} .$$

Let α as in (B.1.1) and (as standard in theory of continued fraction) let

$$q_{-1} := 0 , \quad q_0 := 1 , \quad q_j := q_{j-1} a_j + q_{j-2} , \ \forall j \geq 1 .$$

Let, also,

$$r_j := [a_j; a_{j+1}, a_{j+2}, ...] , \qquad \sigma_j := r_{j+1} + \frac{q_{j-1}}{q_j} .$$

Notice that $r_{j_0+1} = [1; 1^\infty] = 1 + \alpha_{\mathrm{g}}$ and that (compare Appendix 8 of [**24**])

$$|\alpha_{\mathrm{g}} n_1 + n_2| \geq \frac{\gamma_{\mathrm{g}}}{|n_1|} , \qquad \forall n_1 \neq 0$$

where

$$\gamma_{\mathrm{g}} := \frac{2}{3 + \sqrt{5}} .$$

PROPOSITION B.1.1. *Let α, j_0, q_j, σ_j and γ_{g} be as above. For any $n \in \mathbb{Z}^2$ with $n_1 \neq 0$, one has*

$$|\alpha n_1 + n_2| \geq \frac{\bar{\gamma}}{|n_1|} ,$$

[2.1]For generalities, see [**69**].

where

$$\bar{\gamma} := \min_{0 \le j \le j_0} \frac{1}{\sigma_j} \ .$$

For any $n \in \mathbb{Z}^2$ *with* $|n_1| \ge q_{j_0}$, *one has*

$$|\alpha n_1 + n_2| \ge \frac{\gamma_{\rm g}}{|n_1|} \ . \tag{B.1.2}$$

The proof of this result follows easily from the theory of continued fraction; see, e.g., Appendix 8 of [**24**].

B.2. Estimates on $s_{p,k}(\delta)$

In this section we give "accurate" bounds on $s_{p,k}(\delta)$ for $\omega = (\alpha, 1)$ with α quadratic, i.e., numbers for which there exist $\gamma > 0$ and $N \ge 1$ such that

$$|\omega \cdot n| = |\alpha n_1 + n_2| \ge \frac{\gamma}{|n_1|} \ , \qquad \forall \ |n_1| \ge N \ . \tag{B.2.1}$$

The numbers considered in § B.1 are special quadratic numbers.

Recall, now, the definition of $s_{p,k}(\delta)$, which in the present case is:

$$\begin{aligned} s_{p,k}(\delta) \ &:= \ \sup_{n \in \mathbb{Z}^2/\{0\}} \Big(|n_1^{k_1} n_2^{k_2}| \exp(-\delta|n|) |\alpha n_1 + n_2|^{-p} \Big) \ , \qquad k \in \mathbb{N}^2; \\ s_{p,1}(\delta) \ &:= \ \sqrt{s_{p,(1,0)}(\delta)^2 + s_{p,(0,1)}(\delta)^2} \ . \end{aligned} \tag{B.2.2}$$

PROPOSITION B.2.1. *Let* ω, γ *and* N *be as in* (B.2.1) *and let* $s_{p,k}$ *be as in* (B.2.2). *Then, for any* $\delta > 0$ *and any* $p \ge 1$:

$$s_{0,1}(\delta) \ \le \ \frac{\sqrt{2}}{e\,\delta} \ , \tag{B.2.3}$$

$$s_{p,0}(\delta) \ \le \ \max\left\{ 1 \ , \ \left(\frac{p}{e}\right)^p \frac{1}{\gamma^p} \frac{1}{\delta^p} \ , \ \max_{\substack{1 \le n_1 < N \\ 0 \le n_2 < \alpha N + 1}} \frac{\exp(-\delta|n|)}{|\alpha n_1 - n_2|^p} \right\} \ , \tag{B.2.4}$$

$$s_{p,1}(\delta) \ \le \ \sqrt{\hat{s}_{p,e_1}(\delta)^2 + \hat{s}_{p,e_2}(\delta)^2} \ , \tag{B.2.5}$$

where

$$\hat{s}_{p,e_1}(\delta) \ \le \ \max\left\{ \frac{1}{e\,\delta} \ , \ \left(\frac{p+1}{e}\right)^{p+1} \frac{1}{\gamma^p} \frac{1}{\delta^{p+1}} \ , \ \max_{\substack{1 \le n_1 < N \\ 0 \le n_2 < \alpha N + 1}} \frac{n_1 \ \exp(-\delta|n|)}{|\alpha n_1 - n_2|^p} \right\} \ ,$$

$$\hat{s}_{p,e_2}(\delta) \ \le \ \max\left\{ \frac{1}{e\,\delta} \ , \ \frac{2^{\frac{p+1}{2}}}{e} \left(\frac{p}{e}\right)^p \frac{1}{\gamma^p} \frac{1}{\delta^{p+1}} \ , \ \max_{\substack{1 \le n_1 < N \\ 0 \le n_2 < \alpha N + 1}} \frac{n_2 \ \exp(-\delta|n|)}{|\alpha n_1 - n_2|^p} \right\} \ .$$

PROOF. Observing that for any $a > 0$

$$\sup_{t > 0} \Big(t^a \ \exp(-t) \Big) = \left(\frac{a}{e}\right)^a \ ,$$

for any $n \ne 0$ and for any j, one finds

$$|n_j| \ \exp(-\delta|n|) \le |n_j| \ \exp(-\delta|n_j|) = \frac{1}{\delta} \Big(\delta|n_j| \ \exp(-\delta|n_j|) \Big) \le \frac{1}{\delta\,e} \ ,$$

which proves (B.2.3).

To prove (B.2.4), let

$$\alpha_n^p := \alpha_n^p(\delta) := \frac{\exp(-\delta|n|)}{|\omega\cdot n|^p}\ ,$$

so that

$$s_{p,0} = \sup_{n\neq 0}\alpha_n^p = \max\left\{ \sup_{\substack{|n_1|<N\\ |n_2|<\alpha N+1\\ n\neq 0}} \alpha_n^p\ ,\ \sup_{\substack{|n_1|<N\\ |n_2|\geq\alpha N+1}} \alpha_n^p\ ,\ \sup_{|n_1|\geq N}\alpha_n^p \right\}.$$

Now, notice that, if $|n_1| < N$ and $|n_2| \geq \alpha N + 1$, then

$$\sup_{\substack{|n_1|<N\\ |n_2|\geq\alpha N+1}} \alpha_n^p \leq \exp(-\delta|n|) < 1\ .$$

If $|n_1| \geq N$, by (B.2.1),

$$\begin{aligned}\alpha_n^p &\leq \frac{|n_1|^p}{\gamma^p}\ \exp(-\delta|n|) \leq \frac{|n_1|^p}{\gamma^p}\ \exp(-\delta|n_1|)\\ &= \frac{1}{\gamma^p\ \delta^p}\ (\delta|n_1|)^p \exp(-\delta|n_1|)\\ &\leq \frac{1}{\gamma^p\ \delta^p}\left(\frac{p}{e}\right)^p .\end{aligned}$$

This proves (B.2.4).
To prove (B.2.5), let

$$\alpha_{n,j}^p := \alpha_{n,j}^p(\delta) := \frac{|n_j|\ \exp(-\delta|n|)}{|\omega\cdot n|^p}\ .$$

We shall show that

$$\sup_{n\neq 0}\alpha_{n,j}^p \leq \hat{s}_{p,e_j}(\delta)\ , \tag{B.2.6}$$

which, by definition of $s_{p,1}$ yields at once (B.2.5). Mimicking the proof of (B.2.4) one checks easily (B.2.6) for $j = 1$. As for (B.2.6) in the case $j = 2$, we observe that (as above) when $|n_1| < N$ and $|n_2| \geq \alpha N + 1$, then

$$\alpha_{n,2}^p \leq |n_2|\exp(-\delta|n_2|) \leq \frac{1}{\delta\ e}\ ,$$

while if $|n_1| \geq N$, then

$$\begin{aligned}\sup_{|n_1|\geq N}\alpha_{n,2}^p &\leq \frac{1}{\gamma^p}|n_2|\ |n_1|^p\ \exp(-\delta|n|)\\ &\leq \frac{1}{\gamma^p}|n_2|\ |n_1|^p\ \exp\left(-\frac{\delta}{\sqrt{2}}|n_1|\right)\ \exp\left(-\frac{\delta}{\sqrt{2}}|n_2|\right)\\ &\leq \frac{1}{\gamma^p}\left(\frac{\sqrt{2}}{\delta}\right)^{p+1}\left(\frac{\delta|n_1|}{\sqrt{2}}\right)^p \exp\left(-\frac{\delta}{\sqrt{2}}|n_1|\right)\left(\frac{\delta|n_2|}{\sqrt{2}}\right)\ \exp\left(-\frac{\delta}{\sqrt{2}}|n_2|\right)\\ &\leq \frac{2^{\frac{p+1}{2}}}{e}\left(\frac{p}{e}\right)^p\ \frac{1}{\gamma^p}\ \frac{1}{\delta^{p+1}}\ .\end{aligned}$$

The proof of Proposition B.2.1 is completed. □

APPENDIX C

Interval Arithmetic

Real numbers are represented by computers in floating-point notation with a finite number of decimal digits. The number of digits depends on the precision declared in the program code. For example, using Fortran 77 floating-point representation, *single precision* corresponds to approximately seven decimal digits, while *double precision* corresponds to approximately sixteen decimal digits. The rounding-off and propagation errors introduced by the computer can be controlled through the *interval arithmetic* technique, which is briefly recalled in this section (see, e.g., [**48**], [**80**]). Let us precise that our programs run on a *Dec Server 4100 5/466* with operating system *v4.0f* (on which the IEEE standard is implemented).
Only a finite set of numbers (which we denote as the set $\mathcal{R}$ of *representable numbers*) can be exactly represented on a computer (encoded as strings of bits). The idea of interval arithmetic consists in replacing any "real" number r with the smallest interval containing r, say (r_-, r_+), with $r_-, r_+ \in \mathcal{R}$. In order to determine r_- and r_+ we compute as follows two auxiliary quantities, say δ_- and δ_+. Let us write a computer program which performs the following instructions:
• make a cycle on $k \geq 0$, computing $x_k = 1 + \frac{1}{2^k}$, $y_k = x_k - 1$;
• if $y_k = 0$, exit the cycle with index $k = k^*$;
• set

$$\delta_- := 1 - \frac{1}{2^{k^*-1}} , \qquad \delta_+ := 1 + \frac{1}{2^{k^*-1}} .$$

Notice that the condition $y_k = 0$ means that the computer does not *recognize* any further division by 2 of y_k.

Finally, the endpoints of the interval containing r can be defined as

$$r_+ = \begin{cases} r \cdot \delta_+ & \text{if } r \geq 0 \\ r \cdot \delta_- & \text{if } r < 0 \end{cases}$$

$$r_- = \begin{cases} r \cdot \delta_- & \text{if } r \geq 0 \\ r \cdot \delta_+ & \text{if } r < 0 . \end{cases}$$

We take care of exceptional cases by defining a suitable range where all arithmetic operations have a well defined image on the computer. Once the real number r is replaced by the interval (r_-, r_+), any successive operation is performed between intervals according to the following strategy. Let additions, subtractions, multiplications and divisions be denoted as *elementary operations*. We reduce any other operation (including exponentials, logarithms, trigonometric functions, etc.) to a sequence of elementary operations, by computing an approximation, defined as a truncation of the Taylor series expansion, and adding an estimate of the remainder.

Elementary operations between intervals are performed by replacing the exact result with an interval whose endpoints are representable numbers. For example, for any real number c, let us denote by $Up(c)$, $Down(c)$, upper and lower bounds on c, i.e. $c \in (Down(c), Up(c))$. Then, if $a \in (a_-, a_+)$ and $b \in (b_-, b_+)$, where $a, b \in \mathbb{R}$ and $a_\pm, b_\pm \in \mathcal{R}$, the sum between intervals is defined as

$$a + b \in (Down(a_- + b_-), Up(a_+ + b_+)) \equiv (a_-, a_+) + (b_-, b_+) \ .$$

The subtraction between intervals is defined in a similar way. Concerning multiplication and division, several sub-cases must be considered according to the sign of a and b. For completeness, we report the formulae for the multiplication between intervals, say $(c_-, c_+) = (a_-, a_+) \cdot (b_-, b_+)$ (the division will follow similar rules).

If $a_- \geq 0$, then

$$\begin{aligned}
(c_-, c_+) &= (Down(a_- \cdot b_-), Up(a_+ \cdot b_+)) && \text{if } b_- \geq 0 \\
(c_-, c_+) &= (Down(a_+ \cdot b_-), Up(a_- \cdot b_+)) && \text{if } b_+ \leq 0 \\
(c_-, c_+) &= (Down(a_+ \cdot b_-), Up(a_+ \cdot b_+)) && \text{if } b_- < 0 < b_+ \ ;
\end{aligned}$$

If $a_+ \leq 0$, then

$$\begin{aligned}
(c_-, c_+) &= (Down(a_- \cdot b_+), Up(a_+ \cdot b_-)) && \text{if } b_- \geq 0 \\
(c_-, c_+) &= (Down(a_+ \cdot b_+), Up(a_- \cdot b_-)) && \text{if } b_+ \leq 0 \\
(c_-, c_+) &= (Down(a_- \cdot b_+), Up(a_- \cdot b_-)) && \text{if } b_- < 0 < b_+ \ ;
\end{aligned}$$

If $a_- < 0 < a_+$, then

$$\begin{aligned}
(c_-, c_+) &= (Down(a_- \cdot b_+), Up(a_+ \cdot b_+)) && \text{if } b_- \geq 0 \\
(c_-, c_+) &= (Down(a_+ \cdot b_-), Up(a_- \cdot b_-)) && \text{if } b_+ \leq 0 \\
(c_-, c_+) &= \Big(\min\{Down(a_- \cdot b_+), Down(a_+ \cdot b_-)\}, \\
&\qquad \max\{Up(a_- \cdot b_-), Up(a_+ \cdot b_+)\} \Big) && \text{if } b_- < 0 < b_+ \ .
\end{aligned}$$

APPENDIX D

A Guide to the Computer Programs

The results about the stability of the three-body problem have been obtained implementing on the computer the theory developed in the previous chapters. In particular, the computer programs simulate an algebraic manipulator in Fortran 77 and interval arithmetic is performed, when necessary. We have separated the overall implementation of the KAM theory into four main programs, performing different tasks.

i) **Initial data:** Given the initial conditions of the minor body, i.e. the major semi-axis a and the eccentricity e, the first program computes the corresponding frequency and it fixes the *unperturbed* energy E_0 (see § 3.4.1). Let (L_V, G_V) be the values of the action variables, corresponding to the pair (a_V, e_V). The initial conditions for the trapping tori are computed as follows: let $\tilde{L}_\pm := L_V \pm 0.001$ and let $\alpha_\pm := \frac{1}{(\tilde{L}_\pm)^3}$. By means of the continued fraction representation, we modify the frequency, so that it satisfies the Diophantine condition. Precisely, let $\tilde{\alpha}_\pm := [a_0; a_1, a_2, ..., a_N, ...]$ and define the new frequency as $\alpha_\pm := [a_0; a_1, a_2, a_3, a_4, 1^\infty]$. Finally, let $L_\pm := \frac{1}{(\alpha_\pm)^3}$ and $G_\pm := -\frac{1}{(L_\pm)^2} - E_0$. Moreover, the program computes the Diophantine constant associated to $\alpha_\pm$.

ii) **Lindstedt expansion:** Given the initial data as in *i)*, we compute the Lindstedt series expansion of the parametric representation of the invariant torus (see § 4.1). Due to time and memory limitations, the computer is able to calculate the ε-series development up to the order 14, though it was sufficient for us to prove the theorems with an ε-expansion up to the order 12. The computations are performed using quadruple precision, which is a crucial requirement for the estimate of the error terms (see § 4.2).

iii) **Evaluation of the error terms:** We estimate the error functions associated to the approximate solution, constructed as the truncated Lindstedt series expansion as in *ii)* (see § 4.2). Due to the fact that the error terms are relatively small, the use of quadruple precision is extremely important.

iv) **KAM algorithm:** The convergence of the method is obtained through the iteration of the KAM algorithm and the control of the KAM condition (see § 4.3). Double precision is sufficient to perform this task.

We remark that the functions computed by the Lindstedt series expansion have the form:

$$f(x_1, x_2; \varepsilon) := \sum_{j=0}^{N} \varepsilon^j \sum_{n,m \in \mathcal{N}_j} f_{j,nm} \exp\big(i(nx_1 + mx_2)\big) , \tag{D.0.1}$$

where $\mathcal{N}_j$ is a suitable set of indexes in $\mathbb{R}^2$ and $f_{j,nm}$ are real or purely imaginary numbers (depending on whether the function is a sum of cosines or sines). We must recall that in order to apply interval arithmetic, the Fourier coefficients $f_{j,nm}$ are represented as intervals, say $f_{j,nm} = (f^-_{j,nm}, f^+_{j,nm})$. In order to perform operations among functions of the form (D.0.1), we found convenient to adopt the representation based on the following quantities:
• $\texttt{nf}(k)$ $(k = -1, ..., N)$: is an $N+2$-dimensional integer vector, marking the length of the order k of the ε-series expansion. More precisely, having set $\texttt{nf}(-1) = 0$, the order k of the ε expansion runs from $\texttt{nf}(k-1)+1$ to $\texttt{nf}(k)$.
• $\texttt{if}(k, 3)$ $(k = 1, ..., \sum_{j=1}^N |\mathcal{N}_j|)$: is a matrix of integer type with dimension $(\sum_{j=1}^N |\mathcal{N}_j| \times 3)$, such that if $\texttt{if}(k, 1) = 1$, than the series is of sin-type (i.e., the whole series must be multiplied by the imaginary unity), while if $\texttt{if}(k, 1) = 0$ than the series is of cos-type (i.e., no multiplicative factors are present). Moreover, $\texttt{if}(k, 2)$ is the first Fourier index (n in (D.0.1)), while $\texttt{if}(k, 3)$ is the second Fourier index (m in (D.0.1)).
• $(\texttt{cfd}(k), \texttt{cfu}(k))$ is the interval denoting the Fourier coefficient $f_{j,nm}$ (see (D.0.1)). The real vectors $\texttt{cfd}(k)$ and $\texttt{cfu}(k)$ have length $\sum_{j=1}^N |\mathcal{N}_j|$.

Here is the list of programs which can be obtained upon request to one of the authors.

• INDATAPM.FOR: computation of the initial data and of the Diophantine constant for the bounding upper and lower tori. The program runs in quadruple precision (without interval arithmetic).
• PART1.FOR: computation of the Lindstedt series for the functions u, v. Quadruple precision is used and interval arithmetic is implemented.
• PART2.FOR: computation of the error terms, using quadruple precision; interval arithmetic is implemented.
• PART3.FOR: implementation of the KAM algorithm and check of the KAM condition, using double precision; interval arithmetic is implemented.

Bibliography

[1] J. J. Abad, H. Koch, and P. Wittwer, *A renormalization group for Hamiltonians: numerical results*, Nonlinearity **11** (1998), 1185-1194.

[2] M. A. Andreu, A. Celletti, and C. Falcolini, *Break-down of librational invariant surfaces*, Int. J. of Bifurcation and Chaos **9**, 5 (1999), 975-982.

[3] J. H. Applegate, M. R. Douglas, Y. Gursel, G. J. Sussman, and J. Wisdom, *The outer solar system for 210 million years*, Astron. J. **92**, 1 (1986), 176-194.

[4] V. I. Arnold, *Proof of a Theorem by A. N. Kolmogorov on the invariance of quasi-periodic motions under small perturbations of the Hamiltonian*, Russian Math. Survey **18** (1963), 13-40.

[5] V. I. Arnold, *Small divisor problems in classical and Celestial Mechanics*, Russian Math. Survey **18** (1963), 85-191.

[6] V. I. Arnold (editor), *Encyclopedia of Mathematical Sciences*, Dynamical Systems III, Springer-Verlag **3**, 1988.

[7] R. Artuso, G. Casati, and D. L. Shepelyansky, *Breakdown of universality in renormalization dynamics for critical invariant torus*, Europhys. Lett. **15** (1991) 381-386.

[8] R. Artuso, G. Casati, and D. L. Shepelyansky, *Break-up of the spiral mean torus in a volume-preserving map*, Chaos, Solitons & Fractals **2** (1992), 181-190.

[9] D. Bensimon, and L. P. Kadanoff, *Extended chaos and disappearance of KAM trajectories*, Phys. D **13**, 1-2 (1984), 82-89.

[10] A. Berretti, and L. Chierchia, *On the complex analytic structure of the golden invariant curve for the standard map*, Nonlinearity **3**, 1 (1990), 39-44.

[11] A. Berretti, A. Celletti, L. Chierchia, and C. Falcolini, *Natural Boundaries for Area-Preserving Twist Maps*, Journal of Statistical Physics **66** (1992), 1613-1630.

[12] A. Berretti, C. Falcolini, and G. Gentile, *Shape of analyticity domains of the Lindstedt series: The standard map*, Phys. Rev. E **63** (2001), 4 pp.

[13] A. Berretti, and G. Gentile, *Scaling properties for the radius of convergence of a Lindstedt series: the standard map*, Journal de Mathématiques Pures et Appliquées (9) 78, 2 (1999), 159-176.

[14] A. Berretti, and S. Marmi, *Scaling near resonances and complex rotation numbers for the standard map*, Nonlinearity **7** (1994), 603-621.

[15] L. Biasco, L. Chierchia, and E. Valdinoci, *Elliptic two-dimensional invariant tori for the planetary three-body problem*, Arch. Rational Mech. Anal., 170, (2003) 91-135. See also the relative *Corrigendum*, Arch. Rational Mech. Anal., (2006) Digital Object Identifier (DOI) 10.1007/s00205-005-0410-5

[16] L. Biasco, L. Chierchia, and E. Valdinoci, *N-dimensional invariant tori for the planar (N+1)-body problem*, SIAM Journal on Mathematical Analysis, **37**, 5 (2006), 1560-1588

[17] E. M. Bollt, and J. D. Meiss, *Breakup of invariant tori for the four-dimensional semi-standard map*, Physica D **66** (1993), 282-297.

[18] D. Braess, and E. Zehnder, *On the numerical treatment of a small divisor problem*, Numer. Math. **39** (1982), 269-292.

[19] A. Celletti, *Analysis of resonances in the spin-orbit problem in Celestial Mechanics: The synchronous resonance (Part I)*, Journal of Applied Mathematics and Physics (ZAMP) **41** (1990), 174-204.

[20] A. Celletti, *Analysis of resonances in the spin-orbit problem in Celestial Mechanics: Higher order resonances and some numerical experiments (Part II)*, Journal of Applied Mathematics and Physics (ZAMP) **41** (1990), 453-479.

[21] A. Celletti, *Construction of librational invariant tori in the spin-orbit problem*, Journal of Applied Mathematics and Physics (ZAMP) **45** (1993), 61-80.

[22] A. Celletti, and L. Chierchia, *Rigorous estimates for a computer-assisted KAM theory*, J. Math. Phys. **28** (1987), 2078-2086.

[23] A. Celletti, and L. Chierchia, *Construction of analytic KAM surfaces and effective stability bounds*, Commun. in Math. Physics **118** (1988), 119-161.

[24] A. Celletti, and L. Chierchia, *A Constructive Theory of Lagrangian Tori and Computer-assisted Applications*, Dynamics Reported (C.K.R.T. Jones, U. Kirchgraber, H.O. Walther Managing Editors), Springer-Verlag, **4** (New Series) (1995), 60-129.

[25] A. Celletti, and L. Chierchia, *On the stability of realistic three-body problems*, Commun. Math. Phys. **186** (1997), 413-449.

[26] A. Celletti, G. Della Penna, and C. Froeschlé, *Estimate of the transition value of librational invariant curves*, Celestial Mech. Dynam. Astronom. **83**, 1 (2002), 257-274.

[27] A. Celletti, and C. Falcolini, *Construction of invariant tori for the spin-orbit problem in the Mercury-Sun system*, Celestial Mech. Dynam. Astronom. **53** (1992), 113-127.

[28] A. Celletti, and C. Falcolini, *Singularities of periodic orbits near invariant curves*, Physica D **170**, 2 (2002), 87-102.

[29] A. Celletti, C. Falcolini, and Porzio A., *Rigorous numerical stability estimates for the existence of KAM tori in a forced pendulum*, Ann. Inst. H. Poincaré **47** (1987), 85-111.

[30] A. Celletti, and C. Froeschlé, *On the determination of the stochasticity threshold of invariant curves*, Int. J. of Bifurcation and Chaos **5**, 6 (1995), 1713-1719.

[31] A. Celletti, and C. Froeschlé, *Numerical investigation of the break-down threshold for a restricted three-body problem*, Planetary and Space Science **46** (11-12) (1998), 1535-1542.

[32] A. Celletti, C. Froeschlé, and E. Lega, *Determination of the frequency vector in the four dimensional standard mapping*, International Journal of Bifurcation and Chaos **6**, 8 (1996), 1579-1585.

[33] A. Celletti, C. Froeschlé, and E. Lega, *Frequency analysis of the stability of asteroids in the framework of the restricted, three-body problem*, Celestial Mech. Dynam. Astronom. **90**, 3-4 (2004), 245-266

[34] A. Celletti, and A. Giorgilli, *On the numerical optimization of KAM estimates by classical perturbation theory*, Journal of Applied Mathematics and Physics (ZAMP) **39** (1988), 743-747.

[35] A. Celletti, A. Giorgilli, and U. Locatelli, *Improved estimates on the existence of invariant tori for Hamiltonian systems*, Nonlinearity **13** (2000), 2, 397-412.

[36] C. Chandre, M. Govin, H. R. Jauslin, and Koch, H., *Universality for the breakup of invariant tori in Hamiltonian flows*, Physical Review E **57** (1998), 6612-6617.

[37] C. Chandre, and H. R. Jauslin, *Renormalization-group analysis for the transition to chaos in Hamiltonian systems*, Phys. Rev. **365**, 1 (2002), 1-64.

[38] C. Chandre, H. R. Jauslin, and G. Benfatto, *An approximate KAM-renormalization-group scheme for Hamiltonian systems*, J. Stat. Phys. **94** (1999), 241-251.

[39] C. Chandre, H. R. Jauslin, G. Benfatto, and A. Celletti, *Approximate renormalization-group transformation for Hamiltonian systems with three degrees of freedom*, Phys. Rev. E **60**, 5 (1999), 5412-5421.

[40] C. Chandre, J. Laskar, G. Benfatto, and H. R. Jauslin, *Determination of the threshold of the break-up of invariant tori in a class of three frequency Hamiltonian systems*, Physica D **154**, 3-4 (2001), 159-170.

[41] P. J. Channell, and C. Scovel, *Symplectic integration of Hamiltonian systems*, Nonlinearity **3** (1990), no. 2, 231-259.

[42] L. Chierchia, and C. Falcolini, *A direct proof of a theorem by Kolmogorov in Hamiltonian systems*, Ann. Sc. Norm. Su Pisa, Serie IV, **XXI** (1994), 541-593.

[43] L. Chierchia, and C. Falcolini, *Compensations in small divisor problems*, Comm. Math. Phys. **175** (1996), 135-160.

[44] B. V. Chirikov, *A universal instability of many dimensional oscillator systems*, Physics Reports **52** (1979), 264-379.

[45] G. Contopoulos, and N. Voglis,, *A fast method for distinguishing between ordered and chaotic orbits*, Astronomy and Astrophysics **317** (1997), 73-81.

[46] C. Delaunay, *Théorie du Mouvement de la Lune*, Mémoires de l'Académie des Sciences **1**, Tome XXVIII, Paris, 1860.

[47] A. Delshams, and R. de la Llave, *KAM theory and a partial justification of Greene's criterion for nontwist maps*, SIAM J. Math. Anal. **31**, 6 (2000), 1235-1269.

[48] J. P. Eckmann, and P. Wittwer, *Computer methods and Borel sommability applied to Feigenbaum's equation*, Springer Lecture Notes in Physics **227** (1985).

[49] L. H. Eliasson, *Absolutely convergent series expansion for quasi-periodic motions*, MPEJ **2** (1996), 1-33.

[50] D. F. Escande, *Stochasticity in classical Hamiltonian systems: Universal aspects*, Physics Reports **121** (1985), 165-261.

[51] D. F. Escande, and F. Doveil, *Renormalization method for computing the threshold of the large-scale stochastic instability in two degrees of freedom Hamiltonian systems*, J. Stat. Physics **26** (1981), 257-284.

[52] C. Falcolini, and R. de la Llave, *A rigorous partial justification of Greene's criterion*, J. Stat. Phys. **67** (1992), 609-643.

[53] C. Falcolini, and R. de la Llave, *Numerical calculation of domains of analyticity for perturbation theories in the presence of small divisors*, J. Stat. Phys. **67** (1992), 645-666.

[54] J. Féjoz, *Quasiperiodic motions in the planar three-body problem*, J. Differential Equations **183** (2002), no. 2, 303-341.

[55] J. Féjoz, *Démonstration du 'théorème d'Arnold' sur la stabilité du système planétaire (d'après Herman).* (French), Ergodic Theory Dynam. Systems **24** (2004), no. 5, 1521-1582.

[56] C. Froeschlé, and E. Lega, *On the structure of symplectic mappings. The fast Lyapunov indicator: a very sensitive tool. New developments in the dynamics of planetary systems*, Celestial Mech. Dynam. Astronom. **78**, 1-4 (2001), 167-195.

[57] C. Froeschlé, and E. Lega, *On the measure of the structure around the last KAM torus before and after its break-up*, Celestial Mech. Dynam. Astronom.**64** (1996), 21-31.

[58] C. Froeschlé, and E. Lega, *Twist angles: a method for distinguishing islands, tori and weak chaotic orbits. Comparison with other methods of analysis*, Astronomy and Astrophysics **334** (1998), 355-362.

[59] G. Gallavotti, *The Elements of Mechanics*, Springer-Verlag, New York, 1983.

[60] G. Gallavotti, *Twistless KAM tori*, Comm. Math. Phys. **164** (1994), 145-156.

[61] G. Gallavotti, *Twistless KAM tori, quasi flat homoclinic intersections and other cancellations in the perturbation series of certain completely integrable Hamiltonian systems. A review*, Reviews on Mathematical Physics **6** (1994), 343-411.

[62] G. Gallavotti, and G. Gentile, *Majorant series convergence for twistless KAM tori*, Ergodic Theory and Dynamical Systems **15** (1995), 857-869.

[63] G. Gentile, and V. Mastropietro, *Tree expansion and multiscale analysis for KAM tori*, Nonlinearity **8** (1995), 1159-1178.

[64] G. Gentile, and V. Mastropietro, *Methods of analysis of the Lindstedt series for KAM tori and renormalizability in classical mechanics. A review with some applications*, Reviews in Math. Phys **8** (1996), 393-444.

[65] A. Giorgilli, and U. Locatelli, *On classical series expansion for quasi-periodic motions*, MPEJ **3**, 5 (1997), 1-25.

[66] J. M. Greene, *A method for determining a stochastic transition*, J. of Math. Phys. **20** (1979), 1183-1201.

[67] J. M. Greene, *KAM surfaces computed from the Hénon-Heiles Hamiltonian*, Nonlinear dynamics and the beam-beam interaction (Sympos., Brookhaven Nat. Lab., New York, 1979), pp. 257-271, AIP Conf. Proc., 57, Amer. Inst. Physics, New York, 1980.

[68] O. Hald, *On a Newton-Moser type method*, Numer. Math. **23** (1975), 411-426.

[69] G. H. Hardy, and E. M. Wright, *The theory of numbers*, Oxford, 1979.

[70] M. Hénon, *Explorationes numérique du problème restreint IV: Masses egales, orbites non periodique*, Bullettin Astronomique **3**, 1, fasc. 2 (1966), 49-66.

[71] M. Hénon, and C. Heiles, *The applicability of the third integral of motion: Some numerical experiments*, Astron J. **69** (1964), 73-79.

[72] M. Herman, *Sur les courbes invariantes pour les difféomorphismes de l'anneau*, Astérisque **2** (1986), 248 pp.

[73] M. Herman, *Recent results and some open questions on Siegel's linearization theorems of germs of complex analytic diffeomorphisms of $\mathbb{C}^n$ near a fixed point*, VIIIth International Congress on Mathematical Physics (Marseille, 1986), World Sci. Publishing, Singapore, 1987, pp. 138-184.

[74] M. Herman, private communication, 1995

[75] H. R. Jauslin, *Numerical implementation of a K.A.M. algorithm*, Internat. J. Modern Phys. C 4 (1993), no. 2, 317-322.

[76] W. H. Jefferys, and J. Moser, *Quasi-periodic solutions for the three-body problem*, Astronom. J. **7**1 (1966) 568-578.

[77] F. John, *Partial Differential Equations*, Fourth edition. Applied Mathematical Sciences, 1. Springer-Verlag, New York, 1982.

[78] L. Kadanoff, *Scaling for a critical Kolmogorov-Arnold-Moser trajectory*, Phys. Rev. Lett. **47** (1981), 1641-1643.

[79] H. Koch, *A renormalization group for Hamiltonians, with applications to KAM tori*, Erg. Theor. Dyn. Syst. **1**9 (1999), 475-521.

[80] H. Koch, A. Schenkel, and P. Wittwer, *Computer-Assisted Proofs in Analysis and Programming in Logic: A Case Study*, SIAM Rev. **3**8, 4 (1996), 565-604.

[81] A. N. Kolmogorov, *On the conservation of conditionally periodic motions under small perturbation of the Hamiltonian*, Dokl. Akad. Nauk. SSR **9**8 (1954), 527-530.

[82] S. Kurosaki, and Y. Aizawa, *Breakup process and geometrical structure of high-dimensional KAM tori*, Prog. Theor. Phys. **9**8 (1997), 783-793.

[83] O. E. Lanford III, *Computer assisted proofs in analysis*, Physics A **1**24 (1984), 465-470.

[84] O. E. Lanford, *Computer assisted proofs*, in "Computational methods in field theory" (Schladming, 1992), Lecture Notes in Phys. **4**09, Springer, Berlin, 1992, pp. 43-58.

[85] J. Laskar, *Secular evolution of the solar system over 10 million years*, Astronomy and Astrophysics **1**98 (1988), 341-362.

[86] J. Laskar, *The chaotic motion of the solar system. A numerical estimate of the size of the chaotic zones*, Icarus **8**8 (1990), 266-291.

[87] J. Laskar, *Frequency analysis for multi-dimensional systems. Global dynamics and diffusion*, Physica D **6**7 (1993), 257-281.

[88] J. Laskar, *Large scale chaos in the solar system*, Astronomy and Astrophysics **2**87 (1994), L9-L12.

[89] J. Laskar, *Large scale chaos and marginal stability in the solar system*, Celestial Mechanics and Dynamical Astronomy **6**4 (1996), 115-162.

[90] J. Laskar, C. Froeschlé, and A. Celletti, *The measure of chaos by the numerical analysis of the fundamental frequencies. Application to the standard mapping*, Physica D **5**6 (1992), 253-269.

[91] J. Laskar, and P. Robutel, *Stability of the planetary three-body problem. I. Expansion of the planetary Hamiltonian*, Celestial Mech. Dynam. Astronom. **6**2 (1995), 193-217.

[92] C. Liverani, G. Servizi, and G. Turchetti, *Some KAM estimates for the Siegel centre problem*, Lett. Nuovo Cimento (2) **3**9, 17 (1984), 417-423.

[93] C. Liverani, and G. Turchetti, *Improved KAM estimates for the Siegel radius*, J. Statist. Phys. **4**5, 5-6 (1986), 1071-1086.

[94] R. de la Llave, A. Gonzàlez, À. Jorba, and J. Villanueva, *KAM theory without action-angle variables*, Nonlinearity 18 (2005), no. **2**, 855-895.

[95] R. de la Llave, and D. Rana, *Accurate strategies for small divisor problems*, Bull. Amer. Math. Soc. (N. S.) **2**2, 1 (1990), 85-90.

[96] R. de la Llave, and S. Tompaidis, *Computation of domains of analyticity for some perturbative expansions of mechanics*, Physica D **7**1 (1994), 55-81.

[97] R. de la Llave, and S. Tompaidis, *Nature of Singularities for Analyticity Domains of Invariant Curves*, Phys. Rev. Lett. **7**3 (1994), 1459-1462.

[98] R. de la Llave, and S. Tompaidis, *On the singularity structure of invariant curves of symplectic mappings*, Chaos **5**, 1 (1995), 227-237.

[99] U. Locatelli, *Three-body planetary problem: study of KAM stability for the secular part of the Hamiltonian*, Planetary and Space Science **4**6, 11-12 (1998), 1453-1464.

[100] U. Locatelli, C. Froeschlé, E. Lega, and A. Morbidelli, *On the Relationship between the Bruno Function and the Breakdown of Invariant Tori*, Physica D **1**39 (2000), 48-71.

[101] U. Locatelli, and A. Giorgilli, *Invariant tori in the secular motions of the three-body planetary systems*, Celestial Mechanics and Dynamical Astronomy **7**8 (2000), 47-74.

[102] R. S. MacKay, *A renormalization approach to invariant circles in area preserving maps*, Physica D **7** (1983), 283-300.

[103] R. S. MacKay, *Transition to chaos for area-preserving maps*, Lectures Notes in Physics **247** (1985), 390-454.

[104] R. S. McKay, *On Greene's residue criterion*, Nonlinearity **5** (1992), 161-187.

[105] R. S. MacKay, J. D. Meiss, and J. Stark, *An approximate renormalization for the break-up of invariant tori with three frequencies*, Phys. Lett. A **190** (1994), 417-424.

[106] R. S. McKay, and J. Stark, *Locally most robust circles and boundary circles for area-preserving maps*, Nonlinearity **5** (1992), 867-888.

[107] J. Mather, *Nonexistence of invariant circles*, Erg. Theory and Dynam. Systems **4** (1984), 301-309.

[108] A. Mehr, and D. F. Escande, *Destruction of KAM tori in Hamiltonian systems: link with the destabilization of nearby cycles and calculation of residues*, Phys. D **13**, 3 (1984), 302-338.

[109] K. R. Meyer, and D. Schmidt (editors), *Computer Aided Proofs in Analysis*, Springer-Verlag, Berlin, 1991.

[110] J. Moser, *On invariant curves of area-preserving mappings of an annulus*, Nach. Akad. Wiss. Göttingen, Math. Phys. Kl. II **1** (1962), 1-20.

[111] J. Moser, *A rapidly convergent iteration method and non-linear partial differential equations*, Ann. Scuola Norm. Sup. Pisa **20** (1966), 499-535.

[112] J. Moser, *Convergent series expansions for quasi-periodic motions*, Math. Ann. **169** (1967), 136-176.

[113] J. Moser, *Stable and random motions in dynamical systems. With special emphasis on celestial mechanics*, Hermann Weyl Lectures, the Institute for Advanced Study, Princeton, N. J. Annals of Mathematics Studies, No. 77. Princeton University Press, Princeton, N. J.; University of Tokyo Press, Tokyo, 1973.

[114] A. M. Nobili, A. Milani, and M. Carpino, *Fundamental frequencies and small divisors in the orbits of the outer planets*, Astronomy and Astrophysics **210** (1989), 313-336.

[115] A. Olvera, and C. Simó, *An obstruction method for the destruction of invariant curves*, Physica D **26** (1987), 181-192.

[116] I. C. Percival, *Chaotic boundary of a Hamiltonian map*, Physica D **6** (1987), 67-77.

[117] H. Poincarè, *Les Methodes Nouvelles de la Mechanique Celeste*, Gauthier Villars, Paris, 1892.

[118] D. Rana, *Proof of accurate upper and lower bounds to stability domains in small denominator problems*, Ph.D. thesis, Princeton Univ. Press, Princeton, NJ, 1987.

[119] P. Robutel, *Stability of the planetary three-body problem. II. KAM Theory and existence of quasi-periodic motions*, Celestial Mechanics **62** (1995), 219-261.

[120] A. E. Roy, *Orbital Motion*, Adam Hilger Ltd., Bristol, 1978.

[121] H. Rüssmann, *Note on sums containing small divisors*, Comm. Pure Appl. Math. **29**, 6 (1976), 755-758.

[122] H. Rüssmann, *Invariant tori in non-degenerate nearly integrable Hamiltonian systems*, Regul. Chaotic Dyn. **6**, 2 (2001), 119-204.

[123] D. Salamon, and E. Zehnder, *KAM theory in configuration space*, Comment. Math. Helvetici **64** (1989), 84-132.

[124] S. J. Shenker, and L. P. Kadanoff, *Critical behavior of a KAM surface. I. Empirical results*, J. Stat. Phys. **27** (1982), 631-656.

[125] C. L. Siegel, *Iteration of analytic functions*, Ann. Math. **43** (1942), 607-612.

[126] C. L. Siegel, and J. K.Moser, *Lectures on Celestial Mechanics*, Springer-Verlag, Berlin, 1971.

[127] C. Simó, *Effective computations in Hamiltonian dynamics*, Mécanique céleste, 23 pp., SMF Journ. Annu., 1996, Soc. Math. France, Paris, 1996.

[128] G. J. Sussman, and J. Wisdom, *Chaotic evolution of the solar system*, Science **241** (1992), 56-62.

[129] V. Szebehely, *Theory of orbits*, Academic Press, New York and London, 1967.

[130] S. Tompaidis, *Approximation of invariant surfaces by periodic orbits in high-dimensional maps: some rigorous results*, Experiment. Math. **5**, 3 (1996), 197-209.

[131] J.-C. Yoccoz, *Théorème de Siegel, pôlynomes quadratiques et nombres de Brjuno*, Astérisque **231** (1995), 3-88.

[132] E. Zehnder, *Generalized implicit function theorems with applications to some small divisor problems. I*, Comm. Pure Appl. Math. **28** (1975), 91-14.

[133] E. Zehnder, *Generalized implicit function theorems with applications to some small divisor problems. II*, Comm. Pure Appl. Math. **29** (1976), no. 1, 49-111.

[134] Small-Body Orbital Elements, http : //ssd.jpl.nasa.gov/sb_elem.html

Editorial Information

To be published in the *Memoirs*, a paper must be correct, new, nontrivial, and significant. Further, it must be well written and of interest to a substantial number of mathematicians. Piecemeal results, such as an inconclusive step toward an unproved major theorem or a minor variation on a known result, are in general not acceptable for publication.

Papers appearing in *Memoirs* are generally at least 80 and not more than 200 published pages in length. Papers less than 80 or more than 200 published pages require the approval of the Managing Editor of the Transactions/Memoirs Editorial Board.

As of February 28, 2007, the backlog for this journal was approximately 15 volumes. This estimate is the result of dividing the number of manuscripts for this journal in the Providence office that have not yet gone to the printer on the above date by the average number of monographs per volume over the previous twelve months, reduced by the number of volumes published in four months (the time necessary for preparing a volume for the printer). (There are 6 volumes per year, each usually containing at least 4 numbers.)

A Consent to Publish and Copyright Agreement is required before a paper will be published in the *Memoirs*. After a paper is accepted for publication, the Providence office will send a Consent to Publish and Copyright Agreement to all authors of the paper. By submitting a paper to the *Memoirs*, authors certify that the results have not been submitted to nor are they under consideration for publication by another journal, conference proceedings, or similar publication.

Information for Authors

Memoirs are printed from camera copy fully prepared by the author. This means that the finished book will look exactly like the copy submitted.

Initial submission. The AMS uses Centralized Manuscript Processing for initial submissions. Authors should submit a PDF file using the Initial Manuscript Submission form found at `www.ams.org/cgi-bin/peertrack/submission.pl`, or send one copy of the manuscript to the following address: Centralized Manuscript Processing, MEMOIRS OF THE AMS, 201 Charles Street, Providence, RI 02904-2294 USA. If a paper copy is being forwarded to the AMS, indicate that it is for it Memoirs and include the name of the corresponding author, contact information such as email address or mailing address, and the name of an appropriate Editor to review the paper (see the list of Editors below).

The paper must contain a *descriptive title* and an *abstract* that summarizes the article in language suitable for workers in the general field (algebra, analysis, etc.). The *descriptive title* should be short, but informative; useless or vague phrases such as "some remarks about" or "concerning" should be avoided. The *abstract* should be at least one complete sentence, and at most 300 words. Included with the footnotes to the paper should be the 2000 *Mathematics Subject Classification* representing the primary and secondary subjects of the article. The classifications are accessible from `www.ams.org/msc/`. The list of classifications is also available in print starting with the 1999 annual index of *Mathematical Reviews*. The Mathematics Subject Classification footnote may be followed by a list of *key words and phrases* describing the subject matter of the article and taken from it. Journal abbreviations used in bibliographies are listed in the latest *Mathematical Reviews* annual index. The series abbreviations are also accessible from `www.ams.org/publications/`. To help in preparing and verifying references, the AMS offers MR Lookup, a Reference Tool for Linking, at `www.ams.org/mrlookup/`.

Electronically prepared manuscripts. The AMS encourages electronically prepared manuscripts, with a strong preference for $\mathcal{A}_{\mathcal{M}}\mathcal{S}$-LaTeX. To this end, the Society has prepared $\mathcal{A}_{\mathcal{M}}\mathcal{S}$-LaTeX author packages for each AMS publication. Author packages include instructions for preparing electronic manuscripts, samples, and a style file that generates the particular design specifications of that publication series. Though $\mathcal{A}\mathcal{M}\mathcal{S}$-LaTeX is the highly preferred format of TeX, author packages are also available in $\mathcal{A}\mathcal{M}\mathcal{S}$-TeX.

Authors may retrieve an author package from the AMS website starting from `www.ams.org/tex/` or via FTP to `ftp.ams.org` (login as `anonymous`, enter username as password, and type `cd pub/author-info`). The *AMS Author Handbook* and the *Instruction Manual* are available in PDF format following the author packages link from `www.ams.org/tex/`. The author package can also be obtained free of charge by sending email to `tech-support@ams.org` (Internet) or from the Publication Division, American Mathematical Society, 201 Charles St., Providence, RI 02904-2294, USA. When requesting an author package, please specify $\mathcal{A}\mathcal{M}\mathcal{S}$-LaTeX or $\mathcal{A}\mathcal{M}\mathcal{S}$-TeX and the publication in which your paper will appear. Please be sure to include your complete mailing address.

After acceptance. The final version of the electronic file should be sent to the Providence office (this includes any TeX source file, any graphics files, and the DVI or PostScript file) immediately after the paper has been accepted for publication.

Before sending the source file, be sure you have proofread your paper carefully. The files you send must be the EXACT files used to generate the proof copy that was accepted for publication. For all publications, authors are required to send a printed copy of their paper, which exactly matches the copy approved for publication, along with any graphics that will appear in the paper.

Accepted electronically prepared files can be submitted via the web at `www.ams.org/submit-book-journal/`, sent via FTP, or sent on CD-Rom or diskette to the Electronic Prepress Department, American Mathematical Society, 201 Charles Street, Providence, RI 02904-2294 USA. TeX source files, DVI files, and PostScript files can be transferred over the Internet by FTP to the Internet node `ftp.ams.org` (130.44.1.100). When sending a manuscript electronically via CD-Rom or diskette, please be sure to include a message identifying the paper as a Memoir.

Electronically prepared manuscripts can also be sent via email to `pub-submit@ams.org` (Internet). In order to send files via email, they must be encoded properly. (DVI files are binary and PostScript files tend to be very large.)

Electronic graphics. Comprehensive instructions on preparing graphics are available at `www.ams.org/jourhtml/`. A few of the major requirements are given here.

Submit files for graphics as EPS (Encapsulated PostScript) files. This includes graphics originated via a graphics application as well as scanned photographs or other computer-generated images. If this is not possible, TIFF files are acceptable as long as they can be opened in Adobe Photoshop or Illustrator. No matter what method was used to produce the graphic, it is necessary to provide a paper copy to the AMS.

Authors using graphics packages for the creation of electronic art should also avoid the use of any lines thinner than 0.5 points in width. Many graphics packages allow the user to specify a "hairline" for a very thin line. Hairlines often look acceptable when proofed on a typical laser printer. However, when produced on a high-resolution laser imagesetter, hairlines become nearly invisible and will be lost entirely in the final printing process.

Screens should be set to values between 15% and 85%. Screens which fall outside of this range are too light or too dark to print correctly. Variations of screens within a graphic should be no less than 10%.

Inquiries. Any inquiries concerning a paper that has been accepted for publication should be sent to `memo-query@ams.org` or directly to the Electronic Prepress Department, American Mathematical Society, 201 Charles St., Providence, RI 02904-2294 USA.

Editors

This journal is designed particularly for long research papers, normally at least 80 pages in length, and groups of cognate papers in pure and applied mathematics. Papers intended for publication in the *Memoirs* should be addressed to one of the following editors. The AMS uses Centralized Manuscript Processing for initial submissions to AMS journals. Authors should follow instructions listed on the Initial Submission page found at www.ams.org/memo/memosubmit.html.

Algebra to ALEXANDER KLESHCHEV, Department of Mathematics, University of Oregon, Eugene, OR 97403-1222; email: ams@noether.uoregon.edu

Algebra and its application to MINA TEICHER, Emmy Noether Research Institute for Mathematics, Bar-Ilan University, Ramat-Gan 52900, Israel; email: teicher@macs.biu.ac.il

Algebraic geometry to DAN ABRAMOVICH, Department of Mathematics, Brown University, Box 1917, Providence, RI 02912; email: amsedit@math.brown.edu

Algebraic number theory to V. KUMAR MURTY, Department of Mathematics, University of Toronto, 100 St. George Street, Toronto, ON M5S 1A1, Canada; email: murty@math.toronto.edu

Algebraic topology to ALEJANDRO ADEM, Department of Mathematics, University of British Columbia, Room 121, 1984 Mathematics Road, Vancouver, British Columbia, Canada V6T 1Z2; email: adem@math.ubc.ca

Combinatorics to JOHN R. STEMBRIDGE, Department of Mathematics, University of Michigan, Ann Arbor, Michigan 48109-1109; email: FRS@umich.edu

Complex analysis and harmonic analysis to ALEXANDER NAGEL, Department of Mathematics, University of Wisconsin, 480 Lincoln Drive, Madison, WI 53706-1313; email: nagel@math.wisc.edu

Differential geometry and global analysis to LISA C. JEFFREY, Department of Mathematics, University of Toronto, 100 St. George St., Toronto, ON Canada M5S 3G3; email: jeffrey@math.toronto.edu

Dynamical systems and ergodic theory to AMIE WILKINSON, Department of Mathematics, Northwestern University, 2033 Sheridan Road, Evanston, IL 60208-2730; email: transactions@math.northwestern.edu

Functional analysis and operator algebras to DIMITRI SHLYAKHTENKO, Department of Mathematics, University of California, Los Angeles, CA 90095; email: shlyakht@math.ucla.edu

Geometric analysis to WILLIAM P. MINICOZZI II, Department of Mathematics, Johns Hopkins University, 3400 N. Charles St., Baltimore, MD 21218; email: trans@math.jhu.edu

Geometric analysis to MLADEN BESTVINA, Department of Mathematics, University of Utah, 155 South 1400 East, JWB 233, Salt Lake City, Utah 84112-0090; email: bestvina@math.utah.edu

Harmonic analysis, representation theory, and Lie theory to ROBERT J. STANTON, Department of Mathematics, The Ohio State University, 231 West 18th Avenue, Columbus, OH 43210-1174; email: stanton@math.ohio-state.edu

Logic to STEFFEN LEMPP, Department of Mathematics, University of Wisconsin, 480 Lincoln Drive, Madison, Wisconsin 53706-1388; email: lempp@math.wisc.edu

Partial differential equations to GUSTAVO PONCE, Department of Mathematics, South Hall, Room 6607, University of California, Santa Barbara, CA 93106; email: ponce@math.ucsb.edu

Partial differential equations and dynamical systems to PETER POLACIK, School of Mathematics, University of Minnesota, Minneapolis, MN 55455; email: polacik@math.umn.edu

Probability and statistics to KRZYSZTOF BURDZY, Department of Mathematics, University of Washington, Box 354350, Seattle, Washington 98195-4350; email: burdzy@math.washington.edu

Real analysis and partial differential equations to DANIEL TATARU, Department of Mathematics, University of California, Berkeley, Berkeley, CA 94720; email: tataru@math.berkeley.edu

All other communications to the editors should be addressed to the Managing Editor, ROBERT GURALNICK, Department of Mathematics, University of Southern California, Los Angeles, CA 90089-1113; email: guralnic@math.usc.edu.

Titles in This Series

879 **O. García-Prada, P. B. Gothen, and V. Muñoz,** Betti numbers of the moduli space of rank 3 parabolic Higgs bundles, 2007

878 **Alessandra Celletti and Luigi Chierchia,** KAM stability and celestial mechanics, 2007

877 **María J. Carro, José A. Raposo, and Javier Soria,** Recent developments in the theory of Lorentz spaces and weighted inequalities, 2007

876 **Gabriel Debs and Jean Saint Raymond,** Borel liftings of Borel sets: Some decidable and undecidable statements, 2007

875 **C. Krattenthaler and T. Rivoal,** Hypergéométrie et fonction zêta de Riemann, 2007

874 **Sonia Natale,** Semisolvability of semisimple Hopf algebras of low dimension, 2007

873 **A. J. Duncan,** Exponential genus problems in one-relator products of groups, 2007

872 **Anthony V. Geramita, Tadahito Harima, Juan C. Migliore, and Yong Su Shin,** The Hilbert function of a level algebra, 2007

871 **Pascal Auscher,** On necessary and sufficient conditions for L^p-estimates of Riesz transforms associated to elliptic operators on $\mathbb{R}^n$ and related estimates, 2007

870 **Takuro Mochizuki,** Asymptotic behaviour of tame harmonic bundles and an application to pure twistor D-modules, Part 2, 2007

869 **Takuro Mochizuki,** Asymptotic behaviour of tame harmonic bundles and an application to pure twistor D-modules, Part 1, 2007

868 **Gelu Popescu,** Entropy and multivariable interpolation, 2006

867 **Vilmos Totik,** Metric properties of harmonic measures, 2006

866 **William Craig,** Semigroups underlying first-order logic, 2006

865 **Nathanial P. Brown,** Invariant means and finite representation theory of $C*$-algebras, 2006

864 **John M. Lee,** Fredholm operators and Einstein metrics on conformally compact manifolds, 2006

863 **M. Lübke and A. Teleman,** The Universal Kobayashi-Hitchin correspondence on Hermitian manifolds, 2006

862 **Alberto Canonaco,** The Beilinson complex and canonical rings of irregular surfaces, 2006

861 **Leon A. Takhtajan and Lee-Peng Teo,** Weil-Petersson metric on the universal Teichmüller space, 2006

860 **Thomas M. Fiore,** Pseudo limits, biadjoints and pseudo algebras: Categorical foundations of conformal field theory, 2006

859 **N. Arcozzi, R. Rochberg, and E. Sawyer,** Carleson measures and interpolating sequences for Besov spaces on complex balls, 2006

858 **Enrico Valdinoci, Berardino Sciunzi, and Vasile Ovidiu Savin,** Flat level set regularity of p-Laplace phase transitions, 2006

857 **Donatella Danielli, Nocola Garofalo, and Duy-Minh Nhieu,** Non-doubling Ahlfors measures, perimeter measures, and the characterization of the trace spaces of Sobolev functions in Carnot-Carathéodory spaces, 2006

856 **Vladimir Bolotnikov and Harry Dym,** On boundary interpolation for matrix valued Schur functions, 2006

855 **Yevgenia Kashina, Yorck Sommerhäuser, and Yongchang Zhu,** On higher Frobenius-Schur indicators, 2006

854 **Noam Greenberg,** The role of true finiteness in the admissible recursively enumerable degrees, 2006

853 **Joachim Krieger,** Stability of spherically symmetric wave maps, 2006

852 **Viorel Barbu, Irena Lasiecka, and Roberto Triggiani,** Tangential boundary stabilization of Navier-Stokes equations, 2006

851 **Jie Wu,** On maps from loop suspensions to loop spaces and the shuffle relations on the Cohen groups, 2006

850 **Siegfried Echterhoff, S. Kaliszewski, John Quigg, and Iain Raeburn,** A categorical approach to imprimitivity theorems for C^*-dynamical systems, 2006

849 **Katsuhiko Kuribayashi, Mamoru Mimura, and Tetsu Nishimoto,** Twisted tensor products related to the cohomology of the classifying spaces of loop groups, 2006

848 **Bob Oliver,** Equivalences of classifying spaces completed at the prime two, 2006

847 **Eric T. Sawyer and Richard L. Wheeden,** Hölder continuity of weak solutions to subelliptic equations with rough coefficients, 2006

846 **Victor Beresnevich, Detta Dickinson, and Sanju Velani,** Measure theoretic laws for lim–sup sets, 2006

845 **Ehud Friedgut, Vojtech Rödl, Andrzej Ruciński, and Prasad V. Tetali,** A Sharp threshold for random graphs with a monochromatic triangle in every edge coloring, 2006

844 **Amadeu Delshams, Rafael de la Llave, and Tere M. Seara,** A geometric mechanism for diffusion in Hamiltonian systems overcoming the large gap problem: Heuristics and rigorous verification on a model, 2006

843 **Denis V. Osin,** Relatively hyperbolic groups: Intrinsic geometry, algebraic properties, and algorithmic problems, 2006

842 **David P. Blecher and Vrej Zarikian,** The calculus of one-sided M-ideals and multipliers in operator spaces, 2006

841 **Enrique Artal Bartolo, Pierrette Cassou-Noguès, Ignacio Luengo, and Alejandro Melle Hernández,** Quasi-ordinary power series and their zeta functions, 2005

840 **Sławomir Kołodziej,** The complex Monge-Ampère equation and pluripotential theory, 2005

839 **Mihai Ciucu,** A random tiling model for two dimensional electrostatics, 2005

838 **V. Jurdjevic,** Integrable Hamiltonian systems on complex Lie groups, 2005

837 **Joseph A. Ball and Victor Vinnikov,** Lax-Phillips scattering and conservative linear systems: A Cuntz-algebra multidimensional setting, 2005

836 **H. G. Dales and A. T.-M. Lau,** The second duals of Beurling algbras, 2005

835 **Kiyoshi Igusa,** Higher complex torsion and the framing principle, 2005

834 **Kenichi Ohshika,** Kleinian groups which are limits of geometrically finite groups, 2005

833 **Greg Hjorth and Alexander S. Kechris,** Rigidity theorems for actions of product groups and countable Borel equivalence relations, 2005

832 **Lee Klingler and Lawrence S. Levy,** Representation type of commutative Noetherian rings III: Global wildness and tameness, 2005

831 **K. R. Goodearl and F. Wehrung,** The complete dimension theory of partially ordered systems with equivalence and orthogonality, 2005

830 **Jason Fulman, Peter M. Neumann, and Cheryl E. Praeger,** A generating function approach to the enumeration of matrices in classical groups over finite fields, 2005

829 **S. G. Bobkov and B. Zegarlinski,** Entropy bounds and isoperimetry, 2005

828 **Joel Berman and Paweł M. Idziak,** Generative complexity in algebra, 2005

827 **Trevor A. Welsh,** Fermionic expressions for minimal model Virasoro characters, 2005

826 **Guy Métivier and Kevin Zumbrun,** Large viscous boundary layers for noncharacteristic nonlinear hyperbolic problems, 2005

For a complete list of titles in this series, visit the AMS Bookstore at **www.ams.org/bookstore/**.